Three Quarks Missed the Mark

OR: The neutron is not a dud!
(but the Standard Model is!)

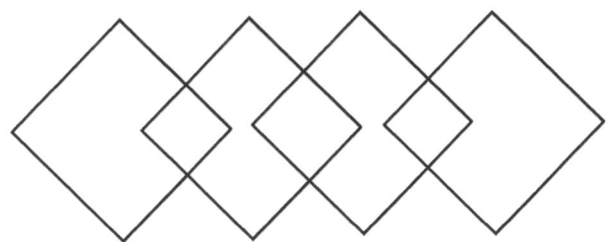

William L. Stubbs

Self-published using
Kindle Direct Publishing
https://kdp.amazon.com/en_US/

ISBN: 9781698130361

Address comments to ift22c@bellsouth.net

For Aunt Kay

Contents

List of Figures

List of Figures (cont.)

Preface

The advent of quantum mechanics in the 1920s and 1930s adversely changed the course of discovery in particle physics. The introduction of uncertainty into physical existence transformed the science from a pursuit of understanding how and why the world works the way it does, into imagining how it might work and relentlessly working to validate those visions. Consequently, a litany of incredible discoveries about the nature and behavior of matter have materialized since that time. The key word here is "incredible." Many of the tenants of modern particle physics are not credible without the assistance of counterintuitive concepts that validate their unusual claims. Enter the Standard Model of Particle Physics.

The Standard Model of Particle Physics is a 17-particle theory of nature that contains particles that cannot be seen in nature, larger particles that spontaneously materialize from smaller particles, particles that can decay into whatever is needed to explain a phenomenon, and particles that can change into other particles and back into their original type while traveling through the vacuum of space. The Standard Model is a collection of lesser theories woven together and presented as a congruent depiction of the fundamental particles.

Three Quarks Missed the Mark examines the Standard Model and shows that deeper analyses reveal that none of its particles are fundamental and some do not even exist. The title is a play on a line from James Joyce's *Finnegans Wake* – "Three quarks for Muster Mark!" – almost always mentioned as where Murray Gell-Mann got the name for the particle from when quarks are introduced. The subtitle – *The neutron is not a dud!* – is also a word play. In the Standard Model, the neutron is made of an up quark (u) and two down quarks (d), making one way to depict it – *dud*.

Bad jokes aside, the book shows that quarks are not the valid interpretation of experimental data, but the blind acceptance of a mathematical device proposed merely to simplify describing and managing subatomic particles. Initially scorned, quarks were quickly embraced once experiments revealed protons are made of smaller particles. Even though very little of the data directly supported the quark hypothesis, it was interpreted in ways that gave life to quarks.

The book shows that, when fully examined, the data points to pions or muons as the likely components of protons, not quarks. Once dismissing the claim that protons contain quarks, *Three Quarks Missed the Mark* goes on to show why, without that, there is no justification for quarks existing in any of the baryons or mesons.

The Standard Model also features leptons – electrons, muons, tau particles and their neutrinos – as fundamental particles. This, even though both the muon and the tau are routinely observed decaying into several other particles. They retain their fundamental status in the Standard model with the help of a couple of other particles – the W and Z bosons.

The W and Z bosons, through their own magic, can come and go as needed, momentarily defying conservation of mass and energy. These extremely massive particles are allowed to be electroweak force carriers along with the photon in the Standard Model, even though they break the symmetry with it because of their mass. This is conveniently circumvented by making the W and Z get their masses from interacting with the Higgs field via the Higgs boson. After nearly 50 years of searching, the Higgs boson was declared found in 2012. This was viewed as the final confirmation of the validity of the Standard Model.

Among the surprises revealed by the book is that free electrons are composite particles. Their complexity is given away by the small difference between their magnetic moment and the Bohr magneton. Also, that there appears to be only one type of neutrino. The neutrinos for the charged leptons all appear to form in the same way, suggesting they are the same particle. Finally, neutrinos appear to be made of negative mass. Perhaps why they are so elusive.

Three Quarks Missed the Mark will likely not change many minds on the nature of matter. Though an attempt to present enough evidence to do so, it is probably too qualitative for hardcore physicists and to quantitative for novices. The hope is that something said in the book will spark a small revolution of thought in the readers' minds that causes them to consider more deeply the plausibility of the Standard Model.

William L. Stubbs
Port St. Lucie, Florida
November 2019

Three Quarks Missed the Mark

1. The Stuff Things Are Made Of

It is likely once people had time enough to think about their surroundings, they wondered what things were made of. As early as the 5th century B.C., philosophers such as Democritus proposed that things were made of indivisible parts he called atoms.[1] While, over time other philosophers, like Aristotle, offered alternatives such as everything made of earth, water, air and fire;[2] atoms prevailed.

By the 17th century Robert Boyle used them to explain the behavior of gases,[3] and in 1800s, John Dalton showed that many substances are made of various groupings of atoms.[4] Dmitri Mendeleev devised the first periodic table of elements in 1871, arranging atoms by their chemical properties.[5]

1.1 Fundamental Particles

Prior to the 20th century, atoms were thought to be the smallest pieces of a material one could get. However, in 1897, J.J. Thompson found that particles eventually called *electrons* came from within atoms, revealing atoms were made of lesser particles.[6] In 1911, Ernest Rutherford showed that, in addition to electrons, atoms have a massive nucleus[7] and in 1919, that it contains *protons*.[8] James Chadwick showed *neutrons* are also part of the nucleus in 1932.[9]

While the atom was being picked apart, others, including James Clerk Maxwell and Albert Einstein, were unravelling the nature of the matter and the forces affecting it. Maxwell laid out the behavior of *electromagnetism* with his famous equations,[10] while Einstein altered our perception of *gravity* with *general relativity* and the relationship between energy and matter with *special relativity*.[11]

In the early 1930s, gravity and electromagnetism were the only two forces thought to exist. However, when neutrons replaced electrons in the nucleus of atoms, it appeared another force was needed to hold the protons in the nucleus against their strong tendency to repel each other. Consequently, the *strong force* was invented. Later, the *weak force* was also invented to explain radioactive decay.

Starting in the 1930s, several new particles smaller than atoms other than electrons, protons and neutrons were found. In 1933, Carl Anderson discovered the positron,[12] a positively charged version of the electron. Seth Neddermeyer led a group in 1937 that discovered the muon,[13] often called a heavy electron. In 1947, Cecil Powell and company discovered the pion,[14] while George Rochester and Clifford Butler found the kaon.[15] The electron neutrino and the muon neutrino were discovered in 1956[16] and 1962,[17] respectively.

In the early 1960s, these discoveries led several physicists including Murray Gell-Mann to show that particles thought to be fundamental could actually be characterized as made of even smaller particles he named quarks.[18,19] Though dismissed as mathematical conjecture at the time;[20] by 1969, a team of scientists using a particle accelerator at Stanford University revealed that protons and neutrons were, indeed, made of small charged particles. Those particles were eventually declared to be the quarks Gell-Mann had proposed.

1.2 The Standard Model

With all the new particles found came a desire to determine those that were fundamental and classify them much like Mendeleev did for the elements with the periodic table.[21,22] In the mid-1970s, physicists dubbed this effort the Standard Model of particle physics.[23,24] The current Standard Model consists of 17 particles divided into two families called fermions and bosons. Fig. 1.1 lays out the 17 particles.

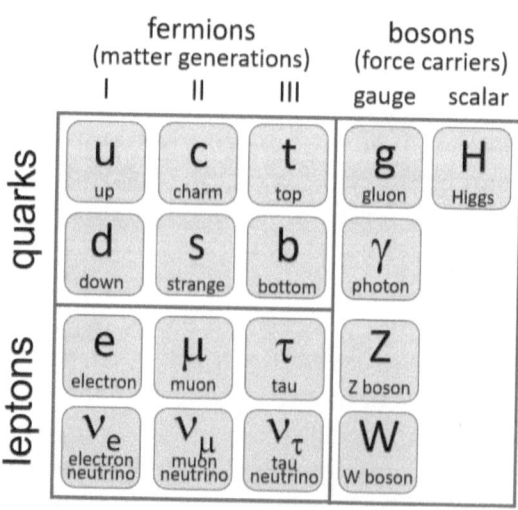

Fig. 1.1: The Standard Model of Particle Physics.

There are 12 fermions, which are mass generators. Six of them are Gell-Mann's quarks and are named: up (*u*), down (*d*), charm (*c*), strange (*s*), top (*t*) and bottom (*b*).[25] The other six, called leptons, are the electron (*e*), the muon (*μ*), the tau (*τ*), the electron neutrino (*v_e*), the muon neutrino (*$v_μ$*) and the tau neutrino (*$v_τ$*).[26] All fermions are charged except the three neutrinos. Each charge fermion has an antiparticle with the same mass but opposite charge of the particle. The antiparticles of the chargeless neutrinos have opposite helicities or handedness of the particles.

The bosons are force carrying particles. They include four vector or gauge bosons: the gluon, the photon, the *Z*, and the *W*; and one scalar boson: the Higgs.[27] The gluon, photon and *Z* are chargeless, but there is a *W+* and a *W-*.

In the Standard Model, combinations of quarks held together by gluons form two classes of particles: baryons and mesons (Fig. 1.2). Baryons are made of three quarks. The only stable baryon is the proton, which is made of two up quarks and one down quark (*uud*). The longest-lived baryon after the proton is the neutron, made of one up quark and two down quarks (*udd*). A free neutrons decays into a proton, an electron and an electron antineutrino in about 13 minutes. All other baryons have lifetimes of less than 10^{-9} seconds.

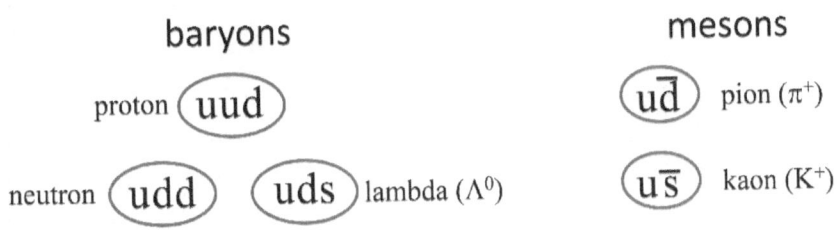

Fig. 1.2: Quark configurations of some baryons and mesons.

The mesons are particles made of two quarks; more specifically, a quark and an antiquark. There are no known stable mesons. The longest-lived mesons, the charged pions (*π+* and *π-*) and kaons (*K+* and *K-*), have lifetimes of just over 10^{-8} seconds. The Standard Model *π+* is an up quark and an anti-down quark (*u$\bar{d}$*), and its *K+* is an up quark and an anti-strange quark (*u$\bar{s}$*). Their antiparticles, *π-* and *K-*, are an antiup quark and a down quark (*$\bar{u}$d*), and an antiup quark and a strange quark (*$\bar{u}$s*), respectively.

In the Standard Model, unstable baryons usually decay into another baryon – either a proton, a neutron or a baryon called a lambda (Λ^0) – and a meson, usually one of the pions. Mesons, other than the pions, typically decay into collections of pions or kaons, while the charged pions decay into leptons and the neutral pion, into photons.

Leptons are another component of matter in addition to the quarks. The only stable charged lepton is the electron. In matter, electrons create the bonds that hold atoms together to form molecules. The muon, which has about 207 times the mass of an electron, eventually decays into an electron and some neutrinos. A free muon lasts for about 10^{-6} seconds. The mass of the tau is about 3,477 times the mass of an electron, making it nearly twice as massive as the proton and the neutron, and more massive than many other baryons.

Because of its large mass, the tau can decay into mesons and the other leptons. Oddly, however, they do not decay into protons, neutrons or other baryons. All tau decay channels ultimately end in the formation of electrons and neutrinos. The lifetime of the tau is about 10^{-13} seconds.

The three neutrinos are all chargeless particles with masses that are thought to be near zero. As such, they do not interact readily with most matter. Their existences were verified through determinations of missing energy and momentum during particle decays. Continuous decay spectrums observed for beta particles emitted during decays of given parent particles hinted at their existences.

The bosons transmit force between the quarks and the leptons in the Standard Model. Photons carry the electromagnetic force between charged particles, gluons transmit the strong force between quarks, and the W and Z bosons produce the weak force. The photon and gluon are massless, but the W has a mass of 80 GeV and the mass of the Z is 91 GeV.

In the Standard Model, the electromagnetic and the weak forces are considered different aspects of the same force. As such, it combines them into one called an electroweak force. This is the first step in combining the four forces into a grand unified theory or GUT.[28]

Presently, the Standard Model does not address transmission of gravity. Particles called gravitons, hypothesized to carry the gravitational force, have yet to be discovered. The discovery of gravity waves in 2016 offers hope that observations of gravitons are coming soon.[29] Efforts are ongoing to roll gravity into Standard Model.

The final particle in the Standard Model is the Higgs boson. Its mass of 125 GeV is about 245,000 times the mass of an electron. Long predicted and final discovered in 2012,[30] the Higgs boson is thought to be the particle evidence of the Higgs field that allegedly gives mass to fundamental particles. Particles that interact with the Higgs boson, such as electrons, have mass, but particles that do not interact with it, such as photons, have no mass. The Higgs boson has a mean lifetime of about 10^{-22} seconds and decays into quarks and W bosons.

1.3 The Baryons Matter

Most elemental matter is made of two baryons, the proton and the neutron. In the Standard Model, the baryons are massive particles made of a collection of three quarks. These are not the only quarks in the baryon, but they are the only ones that do not have an antiparticle mate in it. For that reason, they, at a minimum, determine the type and charge of the baryon. Table 1.1 lists some common baryons and some of their properties.[31]

Table 1.1: Table of Baryons

Particle	Symbol	Quarks	Mass (electrons)	Lifetime (seconds)	Decay Modes
proton	p	uud	1,836.153	stable	none
neutron	n	udd	1,838.684	920	p, e^-, $\underline{v}_e$ (100)
lambda 0	Λ^0	uds	2,183.760	2.6×10^{-10}	p, π^- (63.9) n, π^0 (35.8)
sigma +	Σ^+	uus	2,327.990	0.8×10^{-10}	p, π^0 (51.57) n, π^+ (48.31)
sigma 0	Σ^0	uds	2,334.394	6.0×10^{-20}	Λ^0, γ (100)
sigma -	Σ^-	dds	2,343.803	1.5×10^{-10}	n π^- (100)
delta ++	Δ^{++}	uuu	2,411.431	0.56×10^{-23}	p, π^+ (100)
delta +	Δ^+	uud	2,411.431	0.56×10^{-23}	p, π^0 (100)
delta 0	Δ^0	udd	2,411.431	0.56×10^{-23}	n, π^0 (100)
delta -	Δ^-	ddd	2,411.431	0.56×10^{-23}	n, π^- (100)
xi 0	Ξ^0	uss	2,573.615	2.9×10^{-10}	Λ^0, π^0 (100)
xi -	Ξ^-	dss	2,587.023	1.64×10^{-10}	Λ^0, π^- (100)
omega -	Ω^-	sss	3,273.537	0.82×10^{-10}	Ξ^0, π^- (67.8) Λ^0, K^- (23.6) Ξ^-, π^0 (8.6)

The only stable baryon is the proton. The neutron is essentially stable compared to the other baryons, but still decays into a proton and an electron in about 13 minutes. Consequently, a large portion of the matter seen in the universe is made of baryons.

The other baryons last for less than one-billionth of a second after they form. Being unstable does not eliminate them from being components of stable matter; the unstable neutron is a major part of stable matter. However, to date, none have been found to be part of any enduring matter; and with such brief lifetimes, they likely do not have time to link up with any other particles.

According to the Standard Model, all unstable baryons decay into other baryons and usually a meson, another quark-bearing particle. Two exceptions are shown in the table. The Σ^0 decays into a Λ^0 (another baryon), and a γ, which is a photon; and the neutron decays into a proton and two leptons, the electron and its antineutrino. The diagram in Fig. 1.3 shows the decay channels of the baryons in Table 1.1.

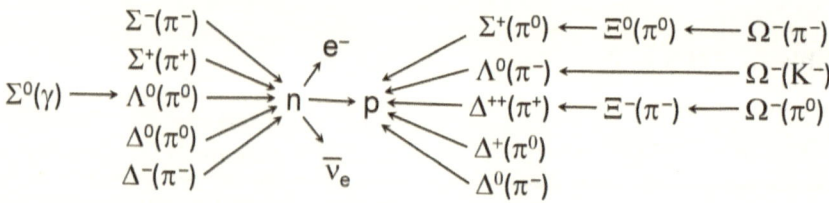

Fig. 1.3: Decay chains of baryons from Table 1.1.

The diagram shows that all the baryons except the Σ^0 and the neutron decay into a different baryon by emitting a meson (shown in parentheses), either a pion or a kaon. All baryons ultimately decay into protons (and electrons). Some decay directly into a proton, some into a neutron, then into a proton, with the Ω^- and the Σ^0, decaying into intermediate baryons before eventually decaying into a proton.

The mesons in the baryon decay channels all eventually decay into electrons and neutrinos. The diagram in Fig. 1.4 shows the paths they take.[32] The charged pions decay into like charged muons and the muons' antiparticle neutrinos. The muons then decay into their neutrinos, like charged electrons and the electrons' antiparticle neutrinos. The neutral pion decays into two photons. The negative kaon can decay into either a neutral pion and a negative pion, which decays into a negative muon; or it can decay directly into a negative muon. The muon decays to an electron and its antineutrino.

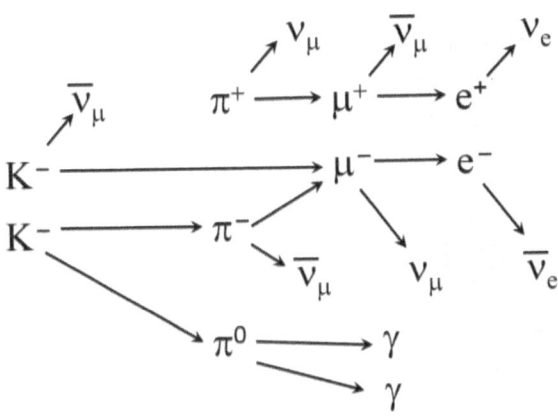

Fig. 1.4: Decay chains of the mesons in Fig. 1.3.

The Standard Model offers two avenues for baryons to decay, one involving the strong interaction and the other, the weak interaction. The strong interaction is the more prevalent one, producing another baryon and a meson as decay products. When it occurs, strangeness, a quantum property of fermions, is conserved.

In the Standard Model, the strangeness number of a fermion is determined by the number of s (strange or anti-strange) quarks it contains. Strangeness is conserved when the decay products contain the same number of strange quarks as the parent fermion. If the parent particle has no s quarks, then the daughters must contain no s quarks. However, if the parent does contain strange quarks, the daughters, together, must have the same number of strange quarks.

For example, the Δ^0 (udd) decays (Fig. 1.5) into a proton (uud) and a π^- ($\bar{u}d$). A gluon, which is a quark-antiquark pair (in this case, $u\bar{u}$), swaps its u quark with a d quark from the Δ^0, converting the Δ^0 to a proton and the gluon to a pion. The gluon makes it a strong interaction. The parent Δ^0 has no strange quarks and daughters, proton and π^-, also have no strange quarks, so strangeness is conserved.

Strangeness conservation is violated during a decay when the number of s quarks in the decay products is different than that of the parent baryon. This can happen when one of the quarks inside a baryon, decays. These decays occur via the weak interaction when a quark emits a W boson, transforming into the next less massive quark.

For example, the Σ⁻ (*dds*) decays (Fig. 1.5) into a neutron (*udd*) and π⁻ (*ūd*). The Σ⁻ has an *s* quark, but neither the neutron nor the π⁻ does. Here, the *s* quark, with a charge of −⅓, emits a *W*⁻, with a charge of −1, becoming a *u* quark, with a charge of +⅔. The *W*⁻ decays into a *ū* quark − *d* quark pair, which is the π⁻ meson.

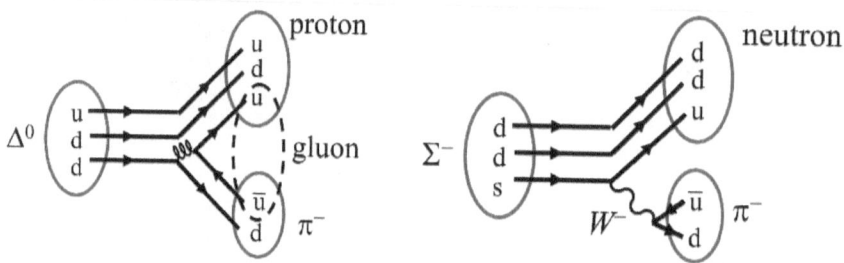

Fig. 1.5: Decay of the Δ⁰ and Σ⁻ baryons.

During Standard Model weak interactions, quarks with negative charges (−⅓) emit *W*⁻ bosons, becoming quarks with +⅔ charges and those with positive charges (+⅔), *W*⁺ bosons, becoming quarks with −⅓ charges. The *W* can then decay into either a pair of quarks or a pair of leptons, depending on the mass of the baryon (Fig. 1.6). The lepton pair is always a particle and its antiparticle neutrino.

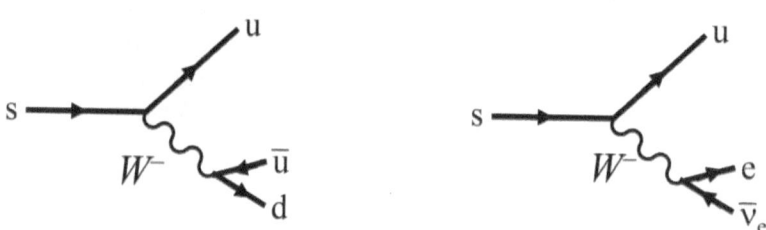

Fig. 1.6: Examples of quark weak interaction decays.

All baryons ultimately decay into protons. The ones more massive than the proton usually result from violent particle collisions (even the neutron). Consequently, these baryons could be viewed as collision fragments that are protons littered with additional mass. Mass that the proton quickly discards as mesons (usually pions) or leptons to cleanse itself. However, the Standard Model does not allow for such a straightforward explanation.

While the quark hypothesis has been crafted into a very efficient tool for modeling and predicting baryon behavior, the fact remains that no quarks have ever been seen. Quantum chromodynamics (QCD) theory was developed to describe why quarks and gluons never appear outside baryons or mesons.[33]

According to QCD, quarks can ever escape the confines of the any baryon or meson as free particles. They are condemned by QCD to always exist in either pairs or triples. Quarks and gluons only exist inside baryons and mesons. Consequently, no free quark or free gluon has ever been directly (or indirectly) observed.

The Standard Model and QCD are based on what the particles inside the proton are believed to be – quarks and gluons, even though they are never seen. Theory to excuse this shortcoming of the model has diverted attention from it; but has not eliminated the problem. Without direct evidence of the quarks, one is still left to wonder what was really found inside the proton at Stanford in the late 1960s?

1.4 They Must Be Quarks

When high-energy electrons (GeV electrons) were fired into the proton at Stanford, they scattered, revealing that the proton has internal structure. The scattering cross sections were used to produce structure function curves that provided details about the proton's structure. Preliminary analyses of the structure functions suggested that the proton is made of charged particles with half-integer spins.

Further analyses were interpreted to indicated that the proton is made of the now familiar quarks and gluons. This model of the proton is very well established, with over 50 years of development and a strong consensus of acceptance as the proper model among the physics community. However, just because there is consensus does not necessarily mean that it is correct. The quark-gluon model of the proton still has the one nagging problem. Though claimed to be resolved via QCD – no quarks or gluons ever appear in nature.

The nonappearance of a hypothesized entity, especially one thought to be as plentiful as the quark, should serve as an indication that the entity probably does not exist. But if protons are not made of quarks and gluons, what then, are they made of? Answering that question requires a reassessment of the electron scattering data that led to the declaration of quarks and gluons as proton components.

2. The Particles Inside the Proton

In the early 1960s, Murray Gell-Mann and others showed that baryons like protons and neutrons could be represented as unitary triplets such as $(b, t, \bar{t})$.[34] Here, the fundamental particles were a neutral baryon b, a singlet s with a charge of z, and a doublet (u, d) with charges $z+1$ and z, respectively. The $\bar{t}$ is the antiparticle of t having the opposite charge of it and z is in units of the absolute value of the electron charge, $|e|$. He considered triplets with spin ½ and $z = -1$, making four particles d^-, s^-, u^0 and b^0 that look like leptons. Then, the proton became $(b, u, \bar{d})$, with a charge of $0 + 0 + -(-1) = +1$.

In 1964, Gell-Mann noticed that a "simpler more elegant scheme" evolved if the fundamental particles carried non-integral charges.[35] When $z = -\frac{1}{3}$, the particles become $d^{-\frac{1}{3}}$, $s^{-\frac{1}{3}}$ and $u^{\frac{2}{3}}$. He dubbed these particles "quarks" (q). With quarks, baryons could be constructed in triplets of (q, q, q) without needing the neutral baryon b. In this scheme, the proton becomes the triplet (u, u, d), which has charge $\frac{2}{3} + \frac{2}{3} + (-\frac{1}{3}) = +1$.

Initially, Gell-Mann appeared to suggest that, with fractional charges, his quarks were merely mathematical entities, not real particles[36] (although later he claimed otherwise[37]). The $z = -\frac{1}{3}$ made the math simpler but did not necessarily imply that particles with this charge or $z = \frac{2}{3}$ really exist in nature. When he introduced quarks in 1964, nearly everyone agreed that quarks could not be real particles.[38]

In 1967, particle physicists from MIT and Stanford began collecting data from the newly built Stanford Linear Accelerator Center (SLAC).[39] The experiments, called deep inelastic scattering, fired extremely high-energy electrons (up to 30 GeV) at stationary proton targets. While Hofstadter had shown that protons are not point particles about a decade earlier,[40] physicists thought them to be homogeneous particles and expected the electrons to pass through them.

To their surprise, the data revealed electrons were scattering at angles indicating the protons have internal structure. Analyses of the data showed that the electrons were scattering off charged particles inside the protons with half-integer spins.[41,42]

2.1 They Could Be Quarks

The discovery of charged particles inside the proton caused physicists to reconsider the prospect that Gell-Mann's quarks might exist. In 1969, Bjorken and Paschos analyzed the scattering data collected to date, intending to show that the proton was, indeed, made of the three quarks Gell-Mann predicted.[43]

Richard Feynman had developed a theory of how particles inside the proton would behave. He suggested that the electrons likely scattered off particles inside the proton he called partons.[44] Similar to Gell-Mann's quarks, Feynman's partons were small, charged particles inside the hadrons. In his model, the electron-proton scattering occurs in the infinite momentum frame of reference.[45]

In the infinite momentum frame, the center-of-mass frame of reference is assumed. There, even though the accelerator fires the electron at the proton, the electron appears to be standing still and the proton moving at near light speed toward the electron. As a result, relativistic time dilation slows down the motion of the particles inside the proton. Then, an impulse approximation[46] is applied to the high-energy collisions between them and the electrons.

The impulse approximation instantaneously frees the individual partons inside the proton from the other particles within the proton. This causes the incident electrons to scatter incoherently off the partons while they are not interacting with other partons. Now, the electron scatters off the partons are elastic, and the scatters give information about the momentum of the individual partons.

While his partons appeared to be similar to Gell-Mann's quarks, it seems Feynman did not initially say that they were or were not quarks. However, upon seeing the results of the initial scattering experiments at SLAC, Feynman did realize that his parton model could explain unexpected scaling behavior observed in the scattering.

Bjorken and Paschos used Feynman's parton theory for their analysis. Scattering cross sections derived from the scattering data were used to produce an F_2 structure function curve.[47] The F_2 values were plotted as a function of the fractions of the proton's momentum, x, particles within the proton struck by electrons carried. In parton theory, F_2 of a proton made of a finite number of particles peaks at the x value that is the reciprocal of the number of particles within the proton. From there, F_2 goes to zero as x approaches zero.

2.2 But They Are Not Quarks

Bjorken and Paschos expected the proton F_2 curve to peak at about $x = \frac{1}{3}$, something like the plot on the top in Fig. 2.1. This would show that the proton was made of three particles, as Gell-Mann predicted. It did not. Instead, the measured data gave the curve on the bottom in Fig. 2.1.

Approaching zero from $x = 1$, the F_2 curve rose until about $x = \frac{1}{5}$ and appeared to remain constant within a wide scattering of data points for the duration of the available data. From this result, Bjorken and Paschos concluded that the proton was not made of just three particles and therefore, not made of Gell-Mann's three quarks.

To salvage Gell-Mann's quarks, Bjorken and Paschos determined that the F_2 values would remain constant as x approached zero only if, in addition to the three quarks, the proton contained an indefinite number of quark-antiquark pairs they called a "pion cloud." Since the quark-antiquark pairs are electrically neutral, three additional quarks are still needed to give the proton an integer charge of +1.

The three quarks became the valence quarks and the quark-antiquark pairs became the sea quarks of the current proton model. After that, the particles inside the proton were always treated as up, down and strange quarks, with the charges $\frac{2}{3}$, $-\frac{1}{3}$ and $-\frac{1}{3}$, respectively. This, even though no particles having charges of $-\frac{1}{3}$ or $+\frac{2}{3}$ had ever been observed in nature, much less, exiting a proton or nucleus.

In 1971, Kuti and Weisskopf showed that adding chargeless gluons to the proton model resolved differences between the measurements and the model in the distribution of momentum among the quarks.[48] Without gluons in the proton model, calculations indicated that the up quarks carried about 18% of the proton's momentum, the down quarks, 6%, and the strange quarks, 76%.

That the strange quarks carried four times the momentum as the up quarks and more than 12 times that of the down quarks seemed out of line with what measurements suggested. When gluons were added to the calculations, the up quarks carried 29% of the momentum, the down quarks carried 19% of the momentum, the strange quarks, 17%, and the gluons, 34% of the momentum. This was considered much more in line with observations. The addition of gluons completed the model of the proton generally recognized today.

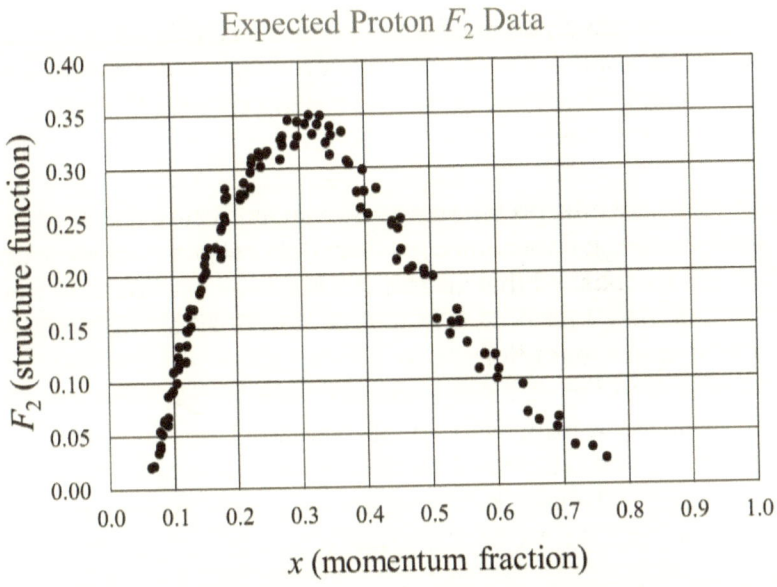

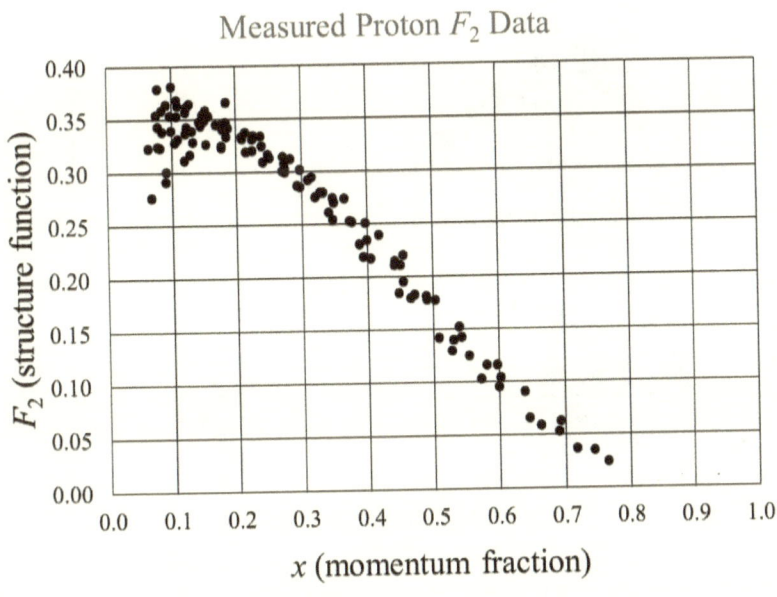

Fig. 2.1: Expected and measured proton F₂ curves.
Type of proton F₂ curve expected for a three-quark proton (top) compared to F₂ curve generated from measured proton data (bottom).

2.3 They Look Like Muons (or Pions?)

There is, however, another way to interpret the SLAC deep inelastic scattering results. This interpretation suggests a proton made of particles readily observed in nature, having charges that are integer multiples of the electron charge. It offers a new paradigm for the structure of the proton that the experimental data also supports.

The alternate interpretation arises from analyzing the SLAC deep inelastic scattering data combined with data collected after the quark model was established that apparently has been overlook. In the following, this new interpretation will be described and contrasted to the current quark paradigm.

The graph in Fig. 2.2 is a composite of the data from the SLAC deep inelastic scattering experiments from the 1960s and experiments performed at the Thomas Jefferson National Accelerator Facility (JLAB) in 1999.[49] The JLAB experiments covered the momentum fraction range from $0 < x < 0.06$ not covered by the SLAC experiments.

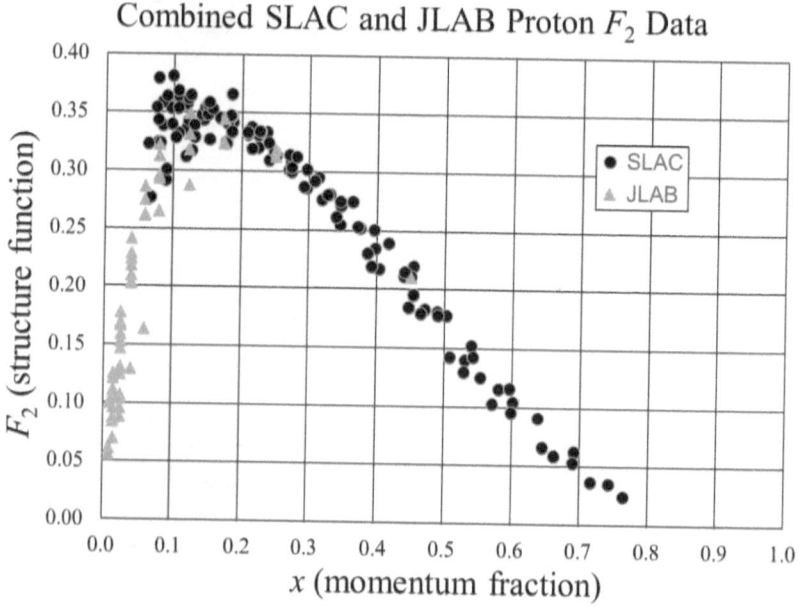

Combined SLAC and JLAB Proton F_2 Data

Fig. 2.2: Combined JLAB and SLAC proton F_2 data.
The combined data shows that the proton F_2 structure function at low-Q^2 peaks at approximately $x = 0.11$. JLAB points at x equals 0.125, 0.25 and 0.45 show that the two curves are comparable.

The range from $0 < x < 0.06$ is that Bjorken and Paschos assumed to stay constant, prompting them to propose the proton model containing valence quarks and sea quarks. In both sets of data displayed, the momentum transferred during the collision between the electron and the proton, Q^2, was less than 3 GeV^2 for x values less than 0.4. This makes the JLAB data comparable to the SLAC data.

The graph shows that, with the JLAB data, the proton F_2 values approach zero as x approaches zero after peaking somewhere between $0.10 \leq x \leq 0.125$. Several of the JLAB points fall within the cluster of SLAC points between $0.06 \leq x \leq 0.20$, and the four points beyond $x = 0.20$ fall well within the scatter of SLAC data in their vicinities. These all show that the JLAB data is comparable to the SLAC data.

This additional data is experimental proof contradicting the assumption made by Bjorken and Paschos in 1969 that the proton F_2 values remain constant for this x region. The assumption they used to justify the existence of three valence quarks and a sea of quark-antiquark pairs as proton components.

According to parton theory, the F_2 curve peaks at the fraction of momentum the particles within the proton carry. This makes it the reciprocal of the number of particles in the proton. The peak F_2 occurring between $0.10 \leq x \leq 0.125$ means that the scattering electrons apparently see between 8 and 10 particles inside the proton.

From the earlier discussions, the particles inside the proton are charged, and they are spin-½ particles. A survey of subatomic particles reveals that the charged, spin-½ particle that is about ⅑ the mass of the proton is the muon.[50] If electrons are finding nine particles inside the proton, one candidate for the particles is the muon.

At a mass of 206.768 electron masses, the free muons are slightly more massive than 204.017 electron masses, one-ninth the proton's mass of 1,836.153 electron masses. Parton theory predicts such a mass difference based on the shape of the proton F_2 curve. The blunt peak of ~ 0.35 ($<< 1.0$) indicates that the particles inside the proton interact strongly, which suggests binding and mass defect.

However, but for the requirement that the particles inside the proton be spin-½ particles, they could also be pions. Pions are slightly more massive than muons, 273 free electron masses versus 207. Therefore, a proton made of pions is likely made of only eight pions, the lower limit prescribed by the electron-proton deep inelastic scattering F_2 analysis.

Eight pions in a proton would have a mass defect of 45 free electron masses per pion, making the binding energy per pion in the proton about 22 MeV. This is nearly the sum of the all the bonds holding an alpha particle together. Being made of eight particles, each with a mass defect of 45 free electron masses means it would take about 178 MeV to completely break a proton into its components.

In contrast, nine free muons, at 207 free electron masses, would only experience a mass defect of about three free electron masses or 1.5 MeV each inside the proton. This would mean that the total proton mass defect would be only 13.5 MeV. This is less than half that of the alpha particle.

Fig. 2.3 shows a photograph of a proton-antiproton collision in a bubble chamber, with a diagram labeling some of the particles evolving in the collision beside it. The diagram shows that emerging from the alleged point of impact are eight tracks claimed to be pions. Four of the tracks curve clockwise, making them negative pions (π^-), and the other four tracks curve counterclockwise, making them positive pions (π^+). This seems to suggest that protons are made of a collection of pions.

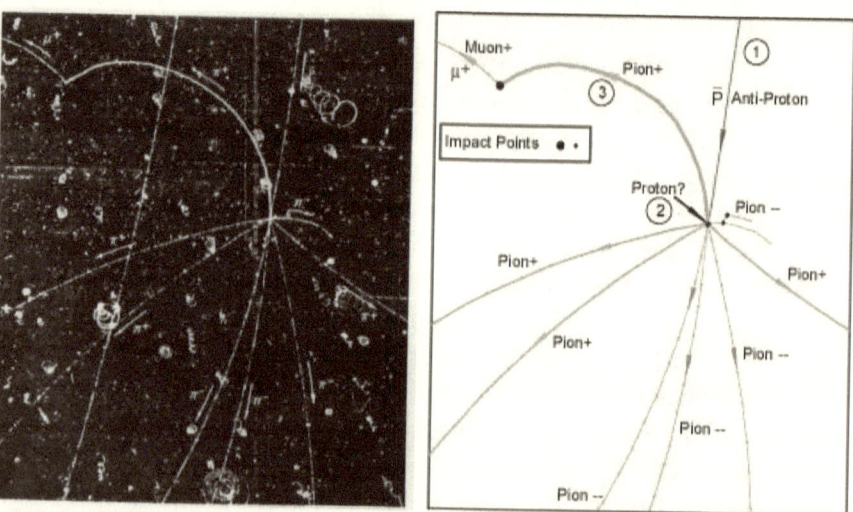

Fig. 2.3: Pions produced in a proton – antiproton collision.
Left: Photograph of a collision of an antiproton with a proton in a bubble chamber.
Right: Diagram of the photo identifying the particles created by the collision. It shows four positive pions and four negative pions formed in the collision. (Lawrence Berkeley National Laboratory Science Photo Library - K003/4377)

The Particles Inside the Proton

The problem with pions as proton components is that they are spin-0 particles. According to an interpretation of some data taken at SLAC, the particles inside the proton are spin-½ particles. The observation that the electrons scattered off particles inside the proton during the scattering is the basis for this conclusion.

The electron-proton deep inelastic scattering is thought to occur because the scattering electron dispatches a virtual photon that is absorbed by the target proton. There is a magnetic interaction between the virtual photon and the proton (F_1 structure function) and an electromagnetic interaction between the two (F_2 structure function).

For the magnetic interaction to occur, the virtual photons must be longitudinal, requiring the target to have a longitudinal cross section, σ_L, to absorb them. The electromagnetic interaction occurs when the photons are transverse, requiring the target to have a transverse cross section, σ_T, to absorb them. The electrons emit virtual photons with helicity ±1, which makes them transverse virtual photons.

Using current algebra, Curtis Callan and David Gross showed that spin-0 particles absorb longitudinal virtual photons, but spin-½ particles absorb transverse virtual photons.[33] They determined that for the particles inside the proton,

$$\omega F_1(\omega) = \frac{1}{4\pi\alpha} \lim_{q2 \to -\infty} q^2 \sigma_T(\omega, q^2), \qquad (2.1)$$

$$F_2(\omega) - \omega F_1(\omega) = \frac{1}{4\pi\alpha} \lim_{q2 \to -\infty} q^2 \sigma_L(\omega, q^2), \qquad (2.2)$$

where $\omega = 2Mx$, q^2 is the negative momentum transfer and $M = 1$, the mass of the proton.

If there are spin-0 particles in the proton, then their $\sigma_L > 0$, but their $\sigma_T = 0$. Therefore, the spin-0 particles cannot absorb the transverse virtual photons emitted by the electrons. From equation (2.1), this makes $F_1 = 0$ for them.

Note, however, since their $\sigma_L > 0$, equation (2.2) shows that the spin-0 particles produce nonzero F_2 values. Therefore, the scatterings still produce information about the particles inside the proton, even though the virtual photon is not absorbed.

If there are spin-½ particles inside the proton, then their $\sigma_T > 0$ and their $\sigma_L = 0$. These particles can absorb the virtual photons emitted by the electrons. Now, equation (2.2) becomes $F_2 = 2xF_1$.

This means that if the measured F_1 and F_2 structure function data satisfies the ratio $2xF_1/F_2 = 1$, then the particles inside the proton are spin-½ particles. However, if $2xF_1/F_2 = 0$ for $x \neq 0$, then F_1 must equal zero and the particles inside the proton are spin-0 particles. If the ratio falls between zero and one, then the virtual photons are apparently encountering a mix of spin-0 and spin-½ particles.

The graph in Fig. 2.4 is a plot of $2xF_1/F_2$ using data from some early SLAC electron-proton deep inelastic scattering experiments. It is done for three momentum transfer (Q^2) ranges and shows that, in all three cases, for $x > 0.25$, the ratio hovers about the value 1.0. This was interpreted to indicate the virtual photons were seeing spin-½ particles inside the proton.

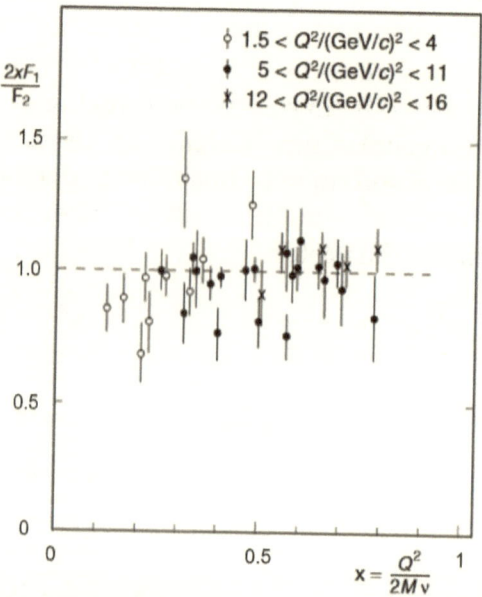

Fig. 2.4: Graph of $2xF_1/F_2$ for electron-proton scattering.
The graph appears to show that for electron-proton scattering, $2xF_1/F_2 = 1$, indicating spin-½ particles inside the proton.

However, for the particles the deep inelastic scattering found inside the proton, $0.10 \leq x \leq 0.125$. For $x < 0.25$, the ratio appears to be steadily declining. This likely indicates that for the smaller x values, the virtual photons are encountering a mix of spin-0 and spin-½ particles. In fact, since the photons probably cannot see the particles inside the proton at $x > 0.25$; there, they are likely responding to the entire proton. The spin-½ they see there is the spin of the proton.

This suggests that, as x gets smaller and the virtual photons focus in on the particles inside the proton, they are encountering spin-0 particles. The virtual photons not being able to see these particles, combined with them still being able to see the entire proton, slowly causes the overall $2xF_1/F_2$ ratio at a given x as $x \to 0$ to get smaller.

Now, the particles inside the proton look like pions, not muons. As a matter of fact, one could probably argue that a line could just as significantly be drawn through a set of points at the ratio value of about 0.8 that spans the x-range of the data. This may be an indication that the virtual photons are seeing spin-0 particles along with the whole proton across the whole span of x-values.

It may be that, in their zeal to see a spin-½ particle, the MIT-SLAC researchers did not question why the low-x values of the ratio on the graph were moving away from 1 as $x \to 0$. Or, maybe it did not occur to them that at higher x values, the virtual photons could be interacting with the whole proton and not the particles within it.

Whatever the reason for this apparent oversight; if valid, it is a severe strike against the concept of a proton made of quarks. If the quarks are spin-½ particles as declared, but the particles found inside the proton are spin-0 particles, then the particles inside the proton cannot be quarks.

A proton made of pions does appear to be consistent with what is seen coming out of other baryons. In chapter 1, essentially all the baryons discussed, the Σ, the Λ, the Δ, the Ξ and the Ω, emit a pion during their decay. Only the neutron and the proton do not release a pion. However, as shown above, protons shatter into pions.

The revelation that protons are made of pions, which means that neutrons are also made of pions, seems to imply that all baryons are made of pions. Pions appear to be structural units of the baryons. That is why they appear in the $\bar{p}p$ collision shown in Fig. 2.3.

Therefore, even though nine muons satisfy the apparent requirement that the proton component particles be spin-½ particles; for now, the component particles of the proton revealed by the deep inelastic scattering are assumed to be eight pions.

Consequently, the JLAB data together with the SLAC data appear to show that the proton is made of eight pions, not three valence quarks and a sea of quark-antiquark pairs as Bjorken and Paschos claimed. Since the proton has a net charge of +1, it is apparently made of positive, negative and neutral pions.

2.4 And Are Made of Smaller Particles

In 1992, the Hadron Electron Ring Accelerator (HERA) produced its first set of electron-proton scattering data.[51, 52] Unlike the linear accelerator in SLAC, which fired electrons at stationary proton targets; in the ring accelerator, both the electrons and the target protons move. This allows it to produce collisions with much higher momentum transfers (Q^2) than the linear accelerator, making it able to resolve much smaller particles.[53]

The second HERA campaign, in 1993,[54] produced electron-proton scattering data for Q^2 from 4.5 to 1600 GeV^2 and x from 0.13 down to 0.000178. At the time (before JLAB), this appeared to fill the gap from $0 < x < 0.06$ left by the SLAC experiments.

The graphs in Fig. 2.5 compare the SLAC, JLAB and HERA F_2 data. The SLAC curve is the result of fitting the 660 data points to a 20-point moving average. The JLAB points are estimated scaling values at given x values from the measured data. The curve is a fourth-order polynomial fit through the points. The HERA points are also estimated scaling values at given x values. The curve through the points is just a smoothed line through points.

The top graph shows that the JLAB data and the HERA data fork at about $x = 0.13$, the peak of the low-Q^2 SLAC-JLAB curve. It seems that from $x = 1$ down to $x = 0.13$, all three experiments see the same thing when the electrons scatter off the proton. However, once the scattering resolves the eight particles inside the proton, the low-Q^2 scatters have seen all they can see. As the momentum fractions, x, approach zero, their wavelengths are too long to resolve anything smaller than those particles.

The high-Q^2 HERA scatters have electrons with much shorter wavelengths. The fact that the F_2 rises beyond $x = 0.13$ as the momentum fraction, x, approaches zero, indicates that those electrons see smaller details within the proton than the low-Q^2 electrons could see.

The bottom graph in Fig. 2.5 is the top graph with a logarithmic momentum fraction axis. It shows that the high-Q^2 HERA data, starting from $x = 0.13$ and approaching zero, behaves like the low-Q^2 SLAC data from $x = 1$, approaching zero. This shows that the HERA electron scattering is resolving particles inside the particles the SLAC-JLAB electrons resolved. The HERA scattering is looking inside the pions that form the proton.

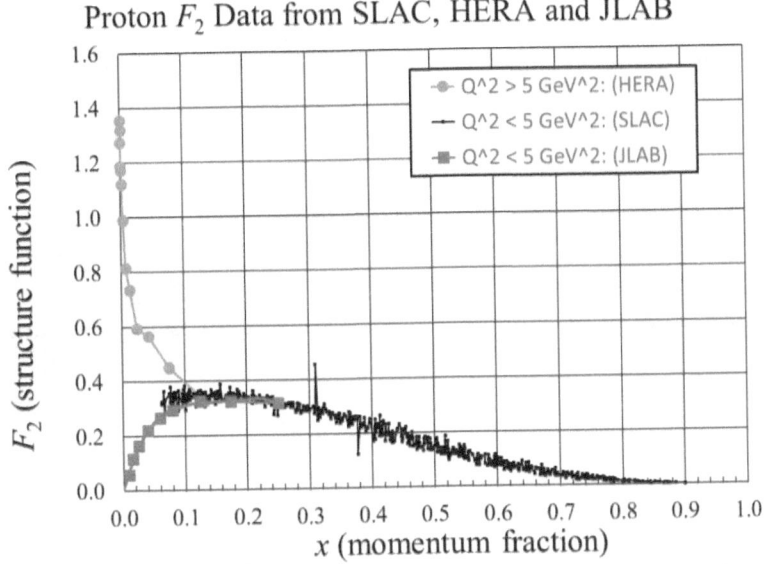

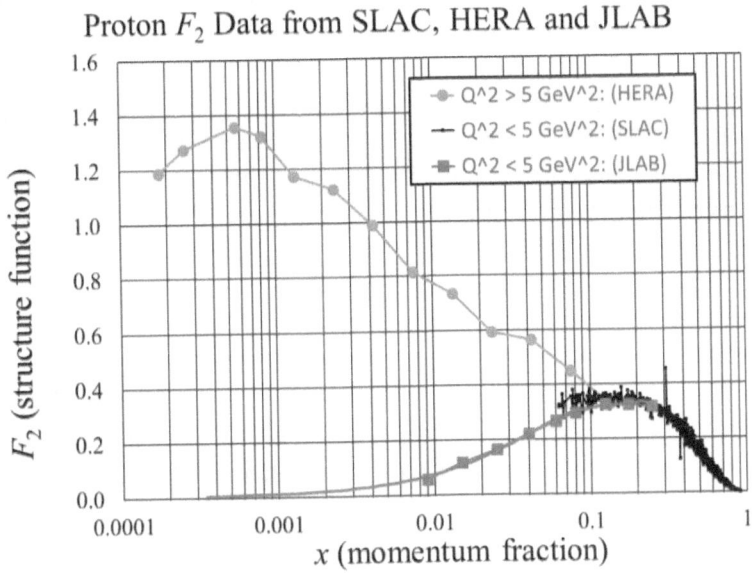

Fig. 2.5: The SLAC, HERA and JLAB proton F₂ data.

The linear (top) and log (bottom) plots of the combined data showing that the proton F_2 structure function forks at about $x = 0.13$. The low-Q^2 JLAB data goes to zero as x approaches zero and the high-Q^2 HERA data rises. The log-version of the graph shows the high-Q^2 data behaves like the low-Q^2 data in a tighter range of x.

The top graph in Fig. 2.6 shows the HERA proton F_2 data for the particles inside the proton with momentum fractions less than 0.125. The graph rises sharply as x approaches zero from 0.125 and peaks near, but at slightly greater than $x = 0$, around $x = 0.0005$. From there, it declines as it continues toward zero.

The sharp peak indicates that the particles that the HERA scattering sees, which are presumably inside the pion, do not interact strongly with each other. This contrasts with the blunt peak the SLAC-JLAB scattering found for the pions inside the proton. They, the pions, apparently interact relatively strongly with each other. They are probably bound to each other like atoms within a molecule.

The particles inside the pions are likely not bound to each other. They are influenced by each other and are probably in orbits or energy levels within the pion like electrons within an atom.

The bottom graph in Fig. 2.6 is the HERA proton F_2 structure function curve (top graph) normalized to a pion F_2 curve. Inspection of the graph reveals that it is the same as the top graph except the values on the axes have changed. This was necessary to make the conversion from the proton curve to the pion curve.

First, the proton is apparently made of eight pions, so each pion carries one-eighth of the proton's momentum. That means that a particle found inside the proton carrying a given fraction of the proton's momentum carries eight times that fraction of the pion's momentum. This makes the proton momentum fraction of 0.125 equal to the pion momentum fraction of 1.0. Consequently, to convert the proton momentum fraction axis to the pion momentum fraction, just multiply its values by eight.

Similarly, the proton F_2 graph shows that at $x = 0.125$, the HERA F_2 value is about $F_2 = 0.35$. This is where the pion momentum fraction is $x = 1.0$. Therefore, by definition, the pion F_2 at this point is $F_2 = 0$. The simplest way to adjust the proton F_2 values to the pion F_2 values is to set $F_2 = 0.35$ for the proton to $F_2 = 0$ for the pion. This is done by subtracting 0.35 from the proton values. This makes the $F_2 = 0.35$ at $x = 0.125$ for the proton, $F_2 = 0$ at $x = 1.0$ for the pion.

The normalized graph shows that when the adjustments are made, the peak F_2 for the particles inside the pion is in the vicinity of $F_2 = 1$. This is another indication that the particles inside the pion are not bound together. The F_2 for completely independent particles would be a δ-function with a spike of $F_2 = 1$.

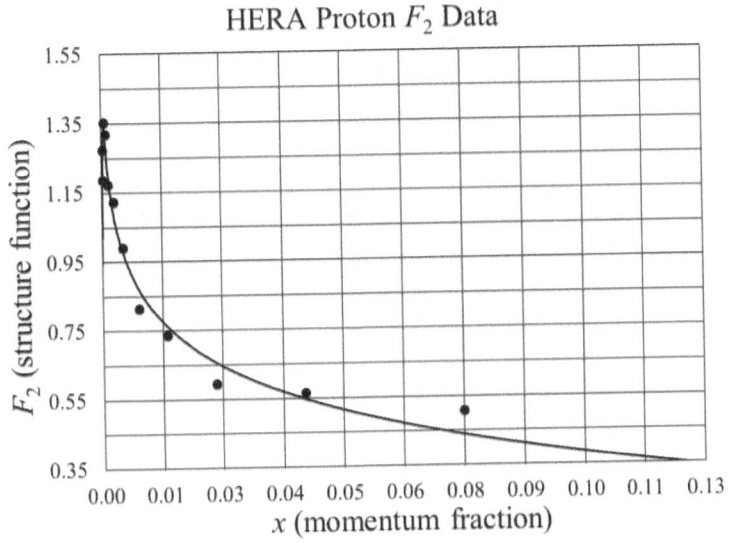

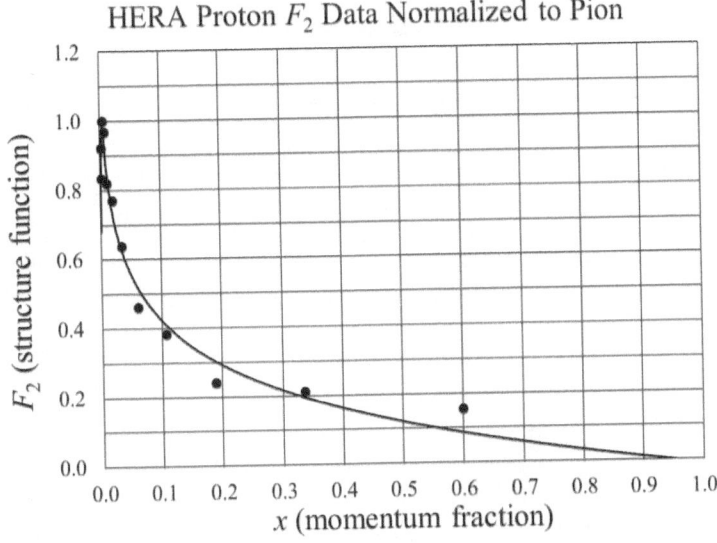

Fig. 2.6: HERA F_2 data for the proton and the pion.
*Top: HERA proton F_2 data plotted for $0 < x < 0.125$. The graph shows the F_2 has a sharp peak near $x = 0$, indicating particles that do not interact strongly with each other. **Bottom:** The HERA proton F_2 curve (top) normalized to a pion F_2 curve. The x-axis is multiplied by 8 to show the fraction of the pion's momentum the particles carry and the F_2 axis has been shifted down by 0.35, the F_2 value where the HERA scattering begins seeing the particles inside the pions.*

2.5 That Look Like Electrons

The graph in Fig. 2.7 shows the pion F_2 curve for pion momentum fractions x between 0 and 0.1. It clearly shows that the pion F_2 values peak at $F_2 \approx 1$ at $x \approx 0.005$. It also shows that the shape of the curve is like that of the proton F_2 curve of the pions. From $x = 0$, it rises to a peak F_2 value, then falls as $x \to 1$. The curve in the figure has been broken into two segments. One segment containing points one through three (triangles), and the other starting at point three and including the remainder of the points (dots). Each set of points has been fitted with a simple logarithmic fit shown on the graph.

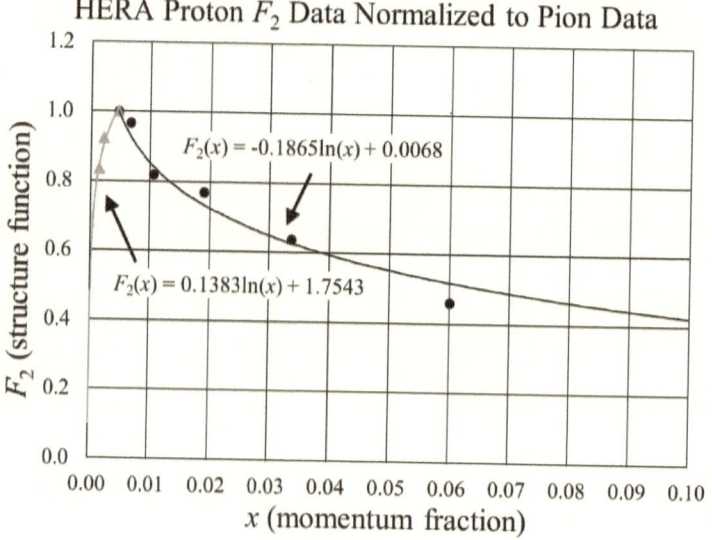

HERA Proton F_2 Data Normalized to Pion Data

$$F_2(x) = -0.1865\ln(x) + 0.0068$$

$$F_2(x) = 0.1383\ln(x) + 1.7543$$

F_2 (structure function)

x (momentum fraction)

Fig. 2.7: HERA pion F_2 data plotted for 0 < x < 0.1.
The pion F_2 curve in two segments. The first (triangles) rises from $F_2 \approx 0$ at $x = 0$ to $F_2 \approx 1$ at $x \approx 0.005$. The second (dots) falls from $F_2 \approx 0.005$ as x rises from 0.005.

The x value where the peak occurs on the normalized F_2 curve indicates the fraction of the pion's momentum its component particles carry. Assuming the two fits should meet at the x-value of the peak F_2, setting the fit equations equal and solving for x should give a good approximate x-value of the peak F_2. The resulting solution is $x = 0.004418$. The reciprocal of this pion x value is 226.3, which means that HERA sees in the neighborhood of 226 particles inside each of the eight pions in the proton.

Eight pions in the proton each having 226 component particles would give the proton 1,808 minor component particles. This is very close to the 1,836 electron masses that makeup the proton. In fact, 1,836 ÷ 8 = 229.5, which means that the pions likely contain an average of 229 particles. The reciprocal of 229 would make $x = 0.004367$ the momentum fraction of the peak F_2 on the pion curve, within just 1.2% of the approximation.

At an average of 229 particles inside the pions in the proton, the pion's components look a lot like electrons (and positrons). For the eight pions of the proton to give it a +1 charge, four could have +1 charges, three, -1, and one, 0. If the four with the +1 charge contain 231 particles, 116 positrons and 115 electrons; the three with the -1 charge contain 231 particles, 115 positrons and 116 electrons; and the neutral one 218 particles, 109 positrons and 109 electrons, the proton would contain 918 positrons and 917 electrons. This would give it 1,835 particles and a charge of +1. The mass of the free pion is about 273 electron masses. If pions are the components of protons, they appear to be made of electrons and positrons.

If this interpretation of the scattering data is valid, then the quark-gluon model of the proton missed this feature of internal proton structure, entirely. The major particles that make up the proton, apparently eight pions, also have structure inside them (Fig. 2.8, left). The quark-gluon model assumes the proton is essentially a container with an unstructured collection of quarks and gluons within it (Fig. 2.8, right). Instead, it appears to have levels of substructure within it.

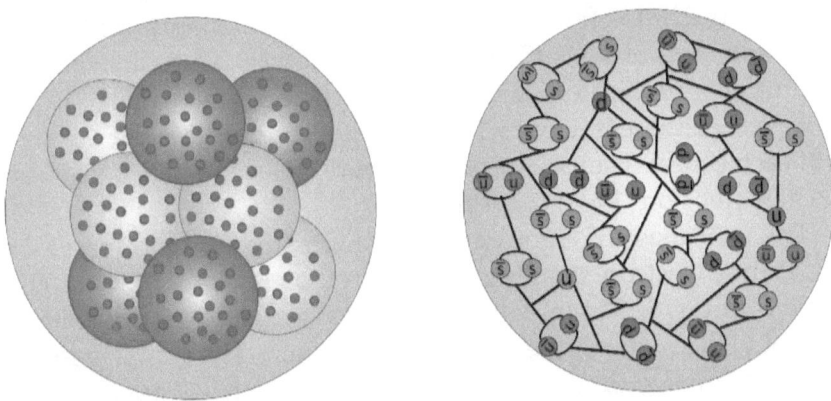

Fig. 2.8: Proton models from scattering data analyses.
Left: Model implied by the reanalysis of the data from SLAC, JLAB and HERA.
Right: Model implied from quark-gluon interpretation of the scattering data.

2.6 A New Model of the Proton

From the experiments performed at SLAC, HERA and JLAB facilities, a clear model of the proton emerges. Critical analyses of their electron-proton deep inelastic scattering data strongly suggest that the proton is likely made of eight pions: possibly four positives, three negatives, each made of about 231 electrons and positrons, and one neutral, pion with 218 electrons and positrons in it. Like the quark-gluon model, a pion-electron model can address why quarks and gluons are never seen leaving the nucleus – because there are none in it!

Unlike quarks and gluons, electrons and positrons are routinely seen exiting the nuclei of many radioactive isotopes. As for pions, physicists have been aware of pions in the debris of proton collisions since the 1950s.[55] Pions are the result of cosmic rays (high-energy protons) colliding with molecules in the Earth's atmosphere[56] and proton-nucleon collisions[57]. It has been known that several pions show up in the debris of proton-proton inelastic scattering collisions since the 1990s.[58]

When a proton comes apart, either pions or electrons and positrons seem to always show up. Consequently, with the pion-electron model of the proton, there is no need for an explanation of why the proton components never appear when a proton is smashed, as is the case for the quark-gluon proton model. The components show up everywhere, all the time.

The low-Q^2 proton F_2 curve indicates that the particles found inside the proton each carry about 12.5% of the proton's momentum. At a mass of 273.132 free electron masses, the pion mass is 0.149 times the proton mass of 1,836.153 free electron masses. The blunt shape of the proton F_2 curve, along with its relatively low peak value of ~ 0.35, indicate that the eight particles inside the proton interact strongly with each other. They are probably bound to each other in clusters, like how nucleons bond to form nuclei.

The total mass of eight pions, 2,185.059 free electron masses, is 348.906 free electron masses greater than the mass of a proton. That converts into 178.291 MeV of mass defect to act as the binding energy that holds the pions inside the proton together. The simplest model of the binding would have the eight pions sharing 349 electrons and positrons between them. That would give each pion a deficit of 45 particles, on average. The pion bonds would be like covalent bonds atoms form in molecules by sharing electrons.

The F_2 curve for the pion indicates that there are likely electrons and positrons contained within it, and that they do not interact strongly with each other. They do not appear to be bound to each other inside the pion like the pions are inside the proton. Instead, since the pion F_2 curve peaks so close to $F_2 = 1$, the electrons and positrons inside it are likely in "orbits" or "shells" within the pion like electrons around the nucleus of an atom.

An orbital configuration can hold the electrons and positrons within the pion without having them cluster together on each other. However, the prospect of this configuration begs the question: If the electrons and positrons making up the pion are in orbits, what are they orbiting? In a later chapter, pion decay suggests that the likely orbital center is a cluster of electrons and positrons.

Finally, eight spherical pions packed tightly within a spherical proton would each have a radius of about 0.378 times the radius of the proton[59]. The charge radius of the proton is about 0.875×10^{-15}m. If the charge is uniformly distributed throughout the proton, then its radius is equal to its charge radius. That would make the radius of the pion about 0.33×10^{-15}m.

3. The Structure of Baryons & Mesons

Eliminating up quarks, down quarks and strange quarks as components of the proton, as well as gluons as its force mediator, has radical implications on the structure of the remaining baryons. If the proton, the only stable baryon, is not made of quarks and gluons, what is the likelihood that the other baryons are made of them?

3.1 Neutron Model Implications

The diagram in Fig. 3.1 shows decay channels of several baryons supposedly made of up, down and strange quarks. All the baryons ultimately decay into a proton, or a neutron, which decays into a proton. Except for the neutron (n) and the neutral sigma (Σ^0), these baryons emit a meson as a decay product.

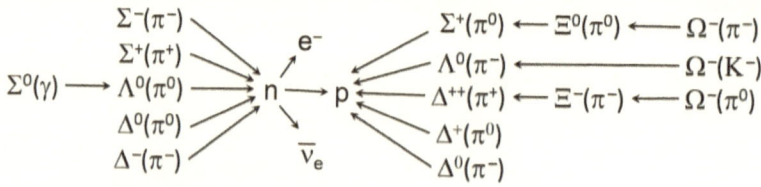

Fig. 3.1: Some baryon decay channels.
The diagram shows how the baryon decay channels all appear to end at the proton.

When the deep inelastic scattering revealed that the proton is made of pions and electrons, by extension, it seems it revealed the neutron is made of pions and electrons, too. With no quarks in the proton, the Standard Model process for converting a neutron made of quarks into a proton cannot occur. In the Standard Model, what starts out as a quark in the neutron must end up a quark in the proton, but it does not.

From its decay, the difference between a neutron and a proton appears to be an electron and an antineutrino. Consequently, the neutron appears to be just a proton with some extra "stuff" in it. Since deep inelastic scattering indicates that there are scores of electrons and positrons within the proton, it is not inconceivable that the neutron is just a proton with some extra electrons and positrons in it.

28

The Standard Model portrays neutron decay as a down quark in the neutron becoming an up quark by emitting a W^- boson via the weak interaction. The diagram in Fig. 3.2 shows how the neutron decay works. A down quark emits a W^-, which changes its charge from $-\frac{1}{3}$ to $+\frac{2}{3}$, making it an up quark. The W^-, which is a virtual particle, exists for only 10^{-25} seconds before it decays into the electron (e^-) – antineutrino $(\bar{v}_e)$ lepton pair.

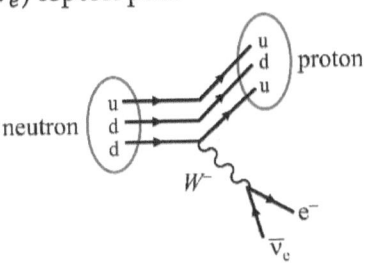

Fig. 3.2: Standard Model decay of the neutron.
A d quark in the neutron emits a W^- boson to become a u quark, making the neutron a proton. The W^- subsequently decays into an electron and an antineutrino.

With no quarks experimentally observed in the proton, the presumption of a down quark in a neutron becoming an up quark to form a proton is invalid. However, neutrons do decay into protons. This means they must do so without the manipulation of quarks by the weak interaction. Therefore, if there are no quarks in the proton, there appear to be no quarks in the neutron, which means all the baryons shown in Fig. 3.1 decay into baryons containing no quarks.

3.2 Baryon Decays

In several instances, a baryon can decay via two channels. For example, the Δ^0 can emit a π^0 to become a neutron or a π^- to become a proton. Similarly, the Σ^+ baryon can emit a π^+ to become a neutron or a π^0 to become a proton. Per the Standard Model, this can occur because the baryons can decay via either the weak interaction or the strong interaction.

As shown in Fig. 3.1, most of the baryons emit a meson (usually a pion) during their decay. Supposedly, this occurs in the Standard Model because baryon masses are so high, they decay via the strong interaction into quark pairs. Except for the neutron, baryons almost always emit mesons, which are quark pairs. Lepton pairs are possible, but most branching fractions for them are less than 10^{-3}.

The diagram on the left of Fig. 3.3 shows an example of a baryon, the Δ^0, decay via the strong interaction into a proton.[60] The Δ^0 supposedly has the same quark configuration as the neutron , *udd*, but has a higher mass. When it decays, like the neutron, a *d* quark in the baryon is converted to a *u* quark.

However, due to the higher mass of the Δ^0, the down quark in the Δ^0 emits an up – anti-up (*uū*) gluon instead of a *W⁻* boson. The *u* quark from the gluon replaces the *d* quark in the baryon, making it a proton. The *ū* quark, pairs with the *d* quark from the original Δ^0 to form a *π⁻* meson.

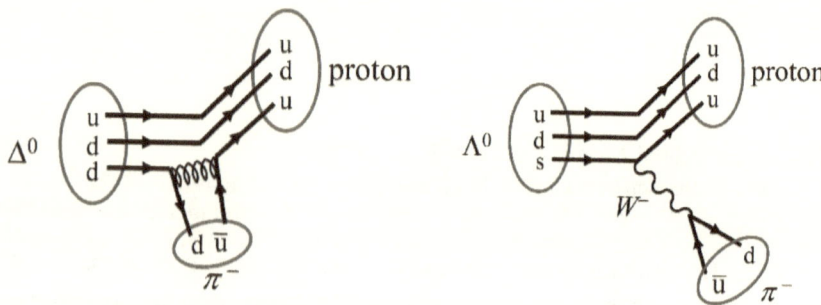

Fig. 3.3: Standard Model decays of the Δ^0 and Λ^0 baryons.
Left: A d quark in the Δ^0 emits a u – ū gluon, which changes the Δ^0 into a proton and creates a π⁻ meson. Right: An s quark in the neutron emits a W⁻ boson to become a u quark, making the Λ^0 a proton. The W⁻ subsequently decays into an anti-up – down quark pair, (ūd), making it a π⁻ meson.

In cases where a strange quark (*s*) is converted to an up quark (*u*), the weak interaction is still thought to be in play. The strong interaction, involving the emission of a gluon, cannot change the strangeness of a baryon. That is, it cannot convert a strange quark with a $-\frac{1}{3}$ charge into an up quark with a $+\frac{2}{3}$ charge. Since the emission of a *W* boson changes the charge of a quark, it can convert a strange quark into an up quark.

For example, Fig. 3.1 shows that the Λ^0 baryon (*uds*) decays into a *π⁻* (*ūd*) and a proton (*uud*).[61] The diagram on the right in Fig. 3.3 shows how the Standard Model proposes this occurs. The *s* quark in the Λ^0 emits a *W⁻*, increasing the charge of the quark from $-\frac{1}{3}$ to $\frac{2}{3}$, making it a *u* quark. This changes the three quarks in the baryon from *uds* to *udu*, making it a Standard Model proton. The *W⁻* decays into an anti-up – down quark pair (*ūd*), which is a *π⁻* meson.

The problem with both these scenarios is, once again, they place a quark in the proton. Since there are no quarks in the proton per the deep inelastic scattering, this process cannot be valid. Now that the proton has been shown not to contain quarks, the conversion of the d and s quarks to u quarks to create the proton is invalid. It cannot happen because there are no quarks in the proton. Therefore, if the Δ^0 and the Λ^0 decay into protons, they must not be made of quarks.

The charmed and bottom baryons also appear to decay into protons and mesons. The charmed baryons Σ_c, Ξ_c, and Ω_c, with known decay channels all decay down to some mesons (pions and/or kaons) and the charmed Λ_c^+ baryon (udc). The Λ_c^+ can decay into either some mesons and a proton (Fig. 3.4, left), or some mesons and a Λ^0 (Fig. 3.4, right), or other less likely combinations that end in protons or Λ^0. The Λ^0 decays into a proton and some mesons (see Fig. 3.3).

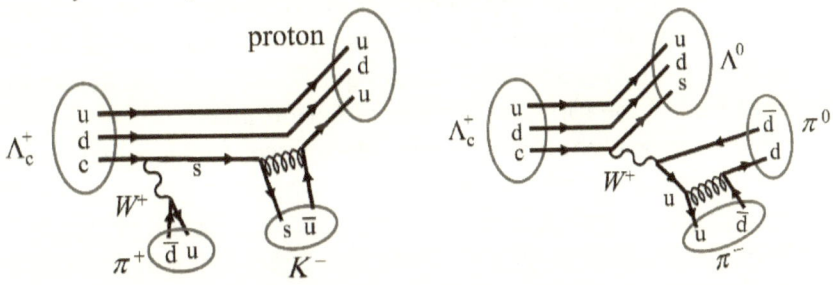

Fig. 3.4: Two decay channels of the Λ_c^+ baryon.
Left: The diagram shows the c quark emitting a W^+ boson, becoming an s quark. The s quark emits a $u\bar{u}$ gluon creating a K^- meson, making the baryon a proton. The W^+ decays into a $u\bar{d}$ pair, which is a π^+ meson. Right: The c quark emits a W^+ that decays into a $u\bar{d}$ (π^+) meson. The u quark in the π^+ emits a $d\bar{d}$ gluon and pairs with the $\bar{d}$ to become a π^- meson. The d from the gluon pairs with the $\bar{d}$ from the W^+ decay to form a π^0 meson.

By Λ_c^+ ultimately decaying into a proton, if charmed baryons contain quarks, their descendants must somehow lose them before becoming protons. The only mechanism the Standard Model appears to provide to change a quark into some particle other than a different quark is annihilation, which makes photons, or W or Z bosons.

With no Standard Model mechanism available to convert the three quarks in its charmed baryons into the eight pions found in protons; the charmed baryons apparently do not contain any quarks. This appears to indicate that the charm quark, c, does not exist.

31

The bottom baryons Σ_b, Ξ_b, and Ω_b, with known decay channels decay down to some mesons and the Λ_b^0 baryon (*udb*). The Λ_b^0 baryon decays by weak interaction into the Λ_c^+ and some mesons and leptons.[62] As shown earlier, the Λ_c^+ decays into a proton.

According to the Standard Model, the bottom baryons ultimately decay into charmed baryons by weak interaction conversion of their bottom quark to a charm quark. It has been determined that the charm quark does not exist as a baryon component if the proton is made of pions and not quarks. Therefore, with no charm quark to decay into, the bottom quark, *b*, must not exist either. Consequently, the bottom baryons must not contain any quarks.

The top quark is thought to have such a short lifetime that it cannot bind with other quarks to form any baryons or mesons. Consequently, there are no baryons or mesons thought to contain top quarks. The top quark supposedly decays into a bottom quark and a W^+ boson. The W^+ decays into leptons. Having concluded that the up, down, strange, charm and bottom quarks do not exist in baryons, the existence of the top quark, *t*, seems very unlikely.

The processes that the Standard Model uses to decay baryons all appear to ultimately rely on the placement of quarks in protons. The electron-proton deep inelastic scattering shows that the proton is not made of quarks, but of pions. Once this is acknowledged, baryons made of quarks are impossible.

3.3 Pions

Analysis of the baryons, against the realization that the proton is not made of quarks, seems to suggest that none of the quarks in the Standard Model actually exist. However, the Standard Model quarks are also found in mesons. It is possible that mesons are made of quarks, but baryons are not.

From the baryon decay channels, it appears that pions are attached to protons and neutrons to form the more complex baryons. For example, about half the time, the Σ^+ baryon decays into a proton and a π^0 meson, and the other half, into a neutron and a π^+ meson. The Σ^- almost always decays into a neutron and a π^- meson.

The lightest Δ^+ baryon, $\Delta(1232)^+$, decays into a proton and a π^0 or a neutron and a π^+, while the heavier $\Delta(1600)^+$ decays into a proton and two pions, a π^+ and a π^-, or a neutron and a π^+ and a π^0. The extra pion appears to contribute to the extra mass of the baryon.

In addition to emitting pions, the heavier charmed and bottom baryons emit heavier mesons, including the K and the D mesons. The D mesons decay into K mesons, which often decay into pions, but can also decay into lepton pairs.

Based on the decays of the heavy baryons and mesons, the pion appears to be a unit of structure in these particles. According to the Standard Model, the π^+ is made of an up quark and an anti-down quark ($u\bar{d}$); the π^-, an anti-up quark and a down quark ($\bar{u}d$); and the π^0, a combination of the up – anti-up ($u\bar{u}$) and the down – anti-down ($d\bar{d}$) quark pairs.

The diagram in Fig. 3.5 shows the decay channels of the mesons produced by the decays of the baryons in Fig. 3.1. It shows that the charged pions decay into like-charged muons and their antineutrinos. This supposedly occurs in the Standard Model because the two quarks making up the pion annihilate each other, forming a like-charged W boson, which decays into the muon and its antineutrino.

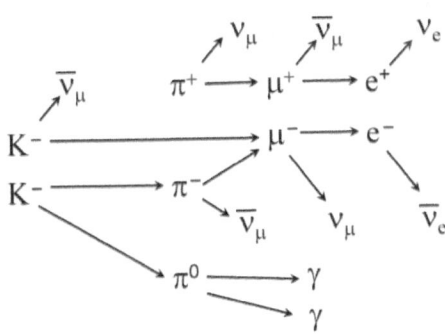

Fig. 3.5: Decay channels of the mesons from Fig. 3.1.
The diagram shows that the charged π and K meson decay channels all appear to end in leptons pairs. The π^0 meson decays to two photons.

The diagram in Fig. 3.6 shows the Standard Model decay of a charged pion. Once again, the Standard Model employs the all-purpose W boson to produce the particles needed to explain an observation. However, a simpler explanation for pion decay is available.

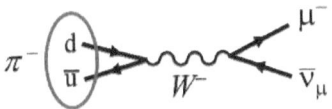

Fig. 3.6: Standard Model decay of π^- to μ^- plus $\bar{\nu}_\mu$.
The $\bar{u}$ and d quarks annihilate forming a W^- that decays into a μ^- and $\bar{\nu}_\mu$.

Electron-proton deep inelastic scattering revealed that the proton is made of pions and that those pions are made of electrons and positrons. Free pions are mesons that decay into muons, which are leptons. Since the pions, with a mass of about 273 free electron masses, appear to be made of about that many electrons and positrons; it does not seem unreasonable that the muons pions decay into are made of about 207 electrons and positrons, their free electron mass equivalent. This makes muons 103 electron-positron pairs and a valence particle.

The charged pion mass of 139.571 MeV is equivalent to 273.133 free electron masses. Consequently, the charged pions may be made of about 273 electrons and positrons. Since the π^- has a charge of -1, it appears to be made of 136 electron-positron pairs and one unpaired electron. The π^+ is also made of 136 electron-positron pairs but has an unpaired positron in it. This model of the charged pions makes them about 33 electron-positron pairs more massive than the muons. Now, when the pions decay into muons, it appears to involve the annihilation of 33 electron-positron pairs creating about 33 MeV.

The mass of the neutral pion, π^0, is 134.977 MeV, or 264.143 free electron masses. This suggests that the π^0 is made of 264 electrons and positrons, 132 pairs. If the decay of the π^0 involves the pairs annihilating, then this model is grossly consistent with observation. When the π^0 meson decays, it produces two photons. There are no massive particles leftover. That means that when the π^0 decays, all 132 electron-positron pairs forming it annihilate. No electrons or positrons are left after the decay and no leptons are formed.

That the π^0, with no net charge on it creates no leptons when it decays provides another insight into the workings of the pions. When the π^- meson decays, apparently 33 of its 136 electron-positron pairs annihilate, and then it becomes a muon (μ^-). The same thing happens with the π^+ meson except that after the 33 electron-positron pairs annihilate, it becomes an antimuon (μ^+).

However, when the electron-positron pairs in the π^0 commence annihilating, they do not stop at 33 pairs. They do not stop, at all. All 132 pairs annihilate. This seems to suggest that the net charge on the charged pions somehow limits the number of electron-positron pairs that can annihilate. Without a valence electron or positron, once the electron-positron pairs in the π^0 start annihilating, there is nothing to stop them. The π^0 flares out.

So, it appears that the net charge on the charged pion is what causes the pion to transform into a muon. It seems that once 33 annihilations have occurred, the charge of the pion makes something happen to form the muon, temporarily stopping the annihilations.

However, the next section on kaons (section 3.4) shows that the much more massive kaon can also decay into a muon and its antineutrino. In doing so, nearly 400 electron-positron pairs annihilate. Therefore, it does not appear to be the number of annihilations that trigger the muon formation, but the number of remaining electron-positron pairs, which is 103 in both the pion and the kaon cases. Apparently, that mass along with the energy from the annihilations and the net charge come together to transform the pion into the muon.

Reviewing the decay channels of the pions reveal that the pions emit neutrinos to become muons. The π^+ emits a muon neutrino to become a μ^+ and the π^-, a muon antineutrino to become a μ^-. The Standard Model interprets the two decay particles as a lepton pair created by the decay of a virtual W boson (see Fig. 3.6). However, this new model of the charged pions can offer a simpler, more insightful explanation of the decay.

When the annihilations occur in the charged pions, once 103 electron-positron pairs remain, the net charge on the pion reacts with its mass and the energy from the annihilations to produce a muon neutrino – muon antineutrino pair. Then, if the net charge of the pion is negative, it retains the muon neutrino to become a μ^-, and the antineutrino zips away. If the charge is positive, the pion holds onto the muon antineutrino, forming a μ^+, and releases the neutrino.

Possessing the muon neutrino is what transforms the pion into a muon. Apparently, the pair production of the neutrinos only temporarily halts the mass annihilation of the electron-positron pairs in the pion turned muon. While the annihilation of the 33 electron-positron pairs in the pions happens in a mean lifetime of 2.6×10^{-8} seconds, the 103 pairs remaining in the newly formed muon annihilate in a mean lifetime of 2.1×10^{-6} seconds. The presence of the muon neutrinos slows down the rate of annihilations, but it does not stop them.

Based on the decay channels of the muons, the charge of the once pion now muon, again reacts with the annihilation energy when only the valence electron or positron is left. However, this time it apparently produces an electron neutrino-antineutrino pair, as signaled by the release of an electron neutrino or antineutrino during the decay.

Also, once all the electron-positron pairs in the muon have annihilated, the neutrino produced by the pion to form the muon is freed. Muon decay is discussed in the next chapter looking at leptons.

This model of the pions supports the electron-proton deep inelastic scattering determination that they are not made of quarks, but of electrons and positrons, like the muons will be shown to be. It provides a simple segue from the pions to the muons and their neutrinos, without the need to conjure up a virtual *W* boson.

The neutrino pair production appears to be insensitive to the polarity of the net charge of the pion, which indicates that some common aspect of positive charge and negative charge must exist. The neutrino pair production happens because the positive and negative charges have a common trait or property that interacts with the pion mass and annihilation energy to form the neutrino matter. However, once the neutrino-antineutrino pair is created, net positive charges seem to attract antineutrinos and repel neutrinos, while negative net charges attract neutrinos and repel antineutrinos.

One can only guess what the physical structure of the pions might be; however, given the significant difference in their lifetimes, the structure of the charged pions is probably different than that of the neutral pion. The relatively long lifetime of the charged pions suggests the electrons and positrons are moving freely within it and annihilating due to chance encounters with an antiparticle (Fig. 3.7).

The short lifetime of the neutral pion coupled with the two gammas formed during its decay point to maybe two, but likely four, subunits within it. Each subunit contains an equal portion of the electrons and positrons. Two subunits collide within the pion, creating a gamma. In a sense, this is similar to the Standard Model pion decay.

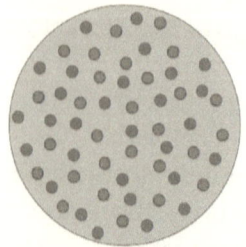 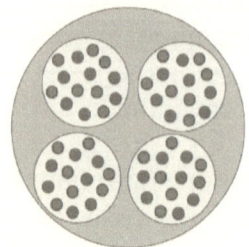

Fig. 3.7: Possible models of the pions
Left: Charged pion with freely moving electrons and positrons. **Right:** *Neutral pion with electrons and positrons divided into four subunits.*

3.4 Kaons (Strange Mesons)

In the Standard Model, the K+ meson is an up quark and an anti-strange quark ($u\bar{s}$) and the K- meson as an anti-up quark and a strange quark ($\bar{u}s$). When charged kaons decay, 63% of the time they decay into a muon and its antineutrino (Fig. 3.8).

Since these daughter products are the same as those of the charged pions, it suggests that the kaon may just be a larger version of the pion. That is, since its mass of 493.677 MeV is equivalent to 966.1 free electron masses, the charged kaons are collections of 967 electrons and positrons, 483 electron-positron pairs and one unpaired electron or positron.

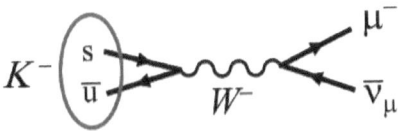

Fig. 3.8: Standard Model decay of K- to μ- plus $\bar{\nu}_\mu$.
The $\bar{u}$ and s quarks annihilate forming a W - that decays into a μ - and $\bar{\nu}_\mu$.

However, 28% of the time, the charged kaons decay into two or three pions. This suggests that the kaon is a collection of as many as three pions. Consequently, it may be that there are several configurations of the charged kaons.

Multiple configurations of the kaon are indicated by the neutral kaon. The mass of the K⁰ meson is 497.611 MeV, which is 973.8 free electron masses. The neutral kaon comes in two varieties: one with a mean lifetime of 0.8954 x 10⁻¹⁰ seconds called K⁰s, the S standing for short (lifetime), and the other, with a mean lifetime of 5.116 x 10⁻⁸ seconds called K⁰L, the L indicating a long lifetime. The significantly different mean lifetimes clearly indicate that the structures of the two are different.

The two types of neutral kaons appear equally when K⁰ mesons are produced. The K⁰s meson almost always (99.89%) decays into two pions, either a π+ and a π- or two π⁰ mesons. It rarely, but can, decay into a π+, a π- and a π⁰ mesons. On the other hand, the K⁰L meson decays into a charged pion and a lepton pair – either an electron or a muon and its antineutrino – two-thirds of the time; and three pions – either three π⁰ mesons or a π+, a π- and a π⁰ mesons – the other third of the time.

If there are multiple configurations of the charged kaon; clearly, the most unstable is the one that is the larger version of the pion. This is likely a spherical cluster of 967 electrons and positrons that looks something like the diagram on the left in Fig. 3.9. In it, pairs of electrons and positrons annihilate until only 103 pairs are left.

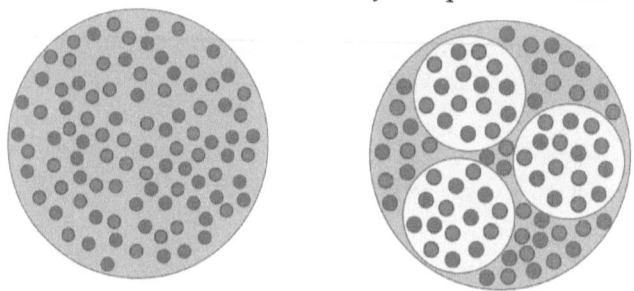

Fig. 3.9: Possible models of the kaons
Left: Charged kaon with freely moving electrons and positrons. Right: Charged kaon made of three pions with electrons and positrons surrounding them.

The mass and net charge of what remains along with the annihilation energy produced then creates a muon neutrino-antineutrino pair, as in the pion. The remaining electrons and positrons capture the appropriate one to turn what is left of the kaon into a muon.

In about 11 out of every 200 charged kaon decays, three charged pions emerge, either π^+, π^+ and π^- for a K$^+$ or π^-, π^- and π^+ for a K$^-$. A Standard Model diagram of the K$^+$ decay is shown in Fig. 3.10. The decay involves five different types of quarks and the emission of a W^+ boson and a gluon, both spontaneously appearing when needed.

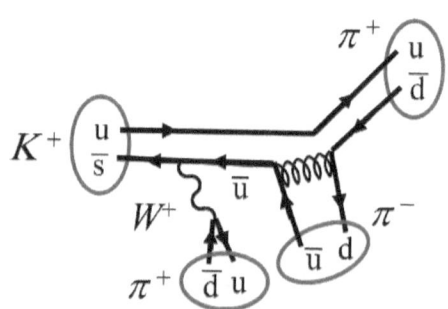

Fig. 3.10: Standard Model decay of K$^+$ to π^+, π^+ and π^-.
The $\bar{s}$ quark emits a W^+, changing it into a $\bar{u}$ quark that emits a $d\bar{d}$ gluon. The $\bar{u}$ pairs with the d from the gluon to form a π^- and the u from the original K$^+$ pairs with the $\bar{d}$ from the gluon to form a π^+. The W^+ decays into a π^+.

As mentioned in the last section, pions appear to be structural units of subatomic particles. The observation of kaons decaying into multiple pions seems to corroborate that notion. This charged kaon might look something like the diagram on the right in Fig. 3.9. There, three pions, a π^+, another π^+ and a π^-, are loosely bound together, moving around within a spherical kaon.

Since the mass of the three pions is just 819 free electron masses, the kaon would have an extra 148 electrons and positrons or 74 electron-positron pairs in it. The pions are embedded in the additional free electrons and positrons. The annihilations of the surrounding electrons and positrons during the decay blast the three pions apart, freeing them.

Though there may be other ways to explain it, these decay channels present a strong case for kaons made of pions. Since the pion has been shown not to be made of quarks, if the charged kaons are made of pions, then they, too, cannot be made of quarks.

Eliminating quarks from the pion challenged the existence of the up quark and the down quark. Showing the kaons are not made of quarks calls into question the existence of the strange quark. Pions and kaons appear to be the components of the charmed and bottom mesons. If so, their presence in these mesons challenge the existence of the charm and bottom quarks.

3.5 Charmed and Bottom Mesons

By the Standard Model, the charmed mesons (D and D_s) are the ones that pair charm (c) quarks and anti-charm ($\bar{c}$) quarks with up quarks, down quarks and strange quarks and their antiquarks. The charged charm mesons D^+ ($c\bar{d}$) and D^- ($\bar{c}d$) have a mass of 1,869.3 MeV, which is 3,658.1 free electron masses. This is about 3.8 times the mass of a charged kaon and 13.4 times the charged pion mass.

The neutral charmed mesons D^0 ($c\bar{u}$) and its antiparticle $\bar{D}^0$ ($\bar{c}u$) have a mass of 1864.5 MeV or 3,648.7 free electron masses, about 10 less than its charged siblings.

The charmed-strange mesons D_s^+ ($c\bar{s}$) and D_s^- ($\bar{c}s$) have a mass of 1,968.5 MeV or 3,852.3 free electron masses, nearly eight times the mass of the charged kaons and 14 times the mass of a the charged pions. There is no neutral D_s meson. All the D mesons or "*deons*" have mean lifetimes on the order of 10^{-12} seconds.

Nearly all the significant decay modes of the deons produce at least one kaon but can emit as many as three. A single kaon is usually accompanied by one, to as many as four, pions of varying polarities. This is an indication that the deons are made of pion and kaon units, which are made of electrons and positrons, not quarks.

The diagram on the left in Fig. 3.11 shows a Standard Model version of the D+ decaying into a K- meson and two π+ mesons. The *c* quark emits a *W+* boson to become an *s* quark. The *s* quark emits a *uū* gluon and takes the *ū* to become a K- meson, while the *u* from the gluon pairs with the *d̄* from the original D+ to form a π+ meson.

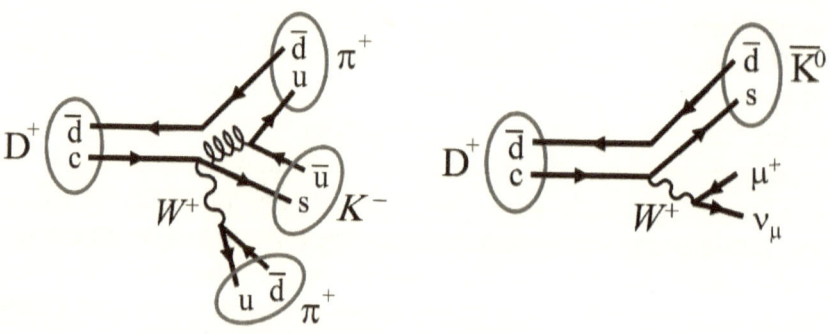

Fig. 3.11: Two Standard Model decays of D+.
Left: The D+ meson emits a W+ boson and a (uū) gluon to decay into two π+ and a K- mesons. Right: The D+ emits a W+ boson to decay into a K̄0 meson and a μ+ - νμ lepton pair.

The lepton pairs $e^+ + \nu_e$, $\mu^+ + \nu_\mu$ and their charge conjugates also appear in some decay channels. The diagram on the right of Fig. 3.11 shows how the D+ decays into the K0 conjugate, a μ+ and a νμ, per the Standard Model. Here, this mode of decay likely signaling that the π and K components of the deons can decay within them.

The D_s mesons are so massive that they can decay into a tau and its neutrino. This suggests that the D_s may have a configuration that is a large spherical collection of 3,850-plus electrons and positrons, a larger version of the simple kaon and pion discussed earlier. The decay channels show that other configurations of the D_s meson can decay into as many as seven pions of various polarities, suggesting that these D_s mesons are just large clusters of pions.[63]

The many decay channels nearly all clearly indicate that the D mesons are built of pion structural units. This means that they are made of electrons and positrons like the proton, not quarks, as the Standard Model professes. Consequently, the charm quark does not appear to exist in these mesons.

In the Standard Model, the bottom mesons or "*beons*" pair the bottom quark (b) with the up, down, strange and charm quarks forming eight particles. The beons are heavy particles. The B$^+$ ($u\bar{b}$), B$^-$ ($\bar{u}b$), B^0 ($d\bar{b}$) and its antiparticle $\bar{B}^0$ ($d\bar{b}$) all have a mass of 5,279 MeV or 10,330.7 free electron masses, about 38 pions or 11 kaons.

The mass of the bottom-charmed mesons B$_c^+$ ($c\bar{b}$) and B$_c^-$ ($\bar{c}b$) is 6,274.9 MeV, which is 12,287.7 free electron masses. This is about 45 charged pions or 13 charged kaons. The neutral bottom-strange meson B$_s^0$ ($s\bar{b}$) and its conjugate B$_s^0$ ($\bar{s}b$) have a mass of 5,366.8 MeV, 10,502.5 free electron masses. It takes about 38 pions or 11 kaons to equal its mass. Like the charmed mesons, the mean lifetimes of the bottom mesons are roughly 10^{-12} seconds.

Like its lighter cousins, the bottom mesons decays into the next lower massive meson more than any other daughter. That is, like K's go to π's and D's go to K's, the B's go to D's. This likely indicates beons are just a continuation of the clumping of less massive mesons, the pions and the kaons, onto large aggregates of structural units. The diagram on the left in Fig. 3.12 shows how a charmed B$_c^+$ emits a W^+ to become a D^0 and a K$^+$ in the Standard Model.[64]

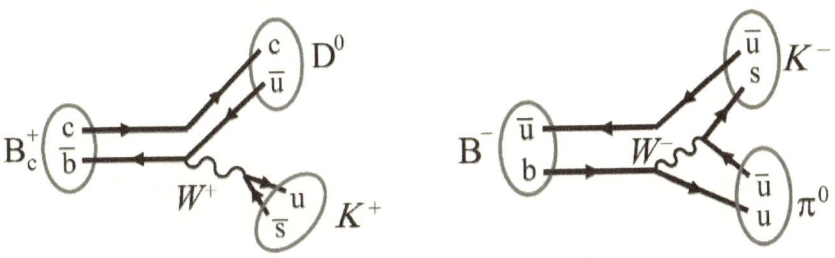

Fig. 3.12: Standard Model decays of the B$^+$ and the B$^-$ mesons
Left: The B$^+$ meson emits a W^+ boson to decay into a D^0 meson and a K$^+$ mesons. Right: The B$^-$ emits a W^- boson to decay into a K$^-$ meson and a π^0 meson.

In the Standard Model, a b quark of the B$^-$ meson can emit a W^- boson that splits it into a K$^-$ meson and a π^0 meson, as shown on the right in Fig. 3.12. If the B$^-$ is made of π mesons and K mesons, then these mesons may survive and escape during its decay.

In a twist not seen in the other mesons, the beons can decay into baryons, including protons.[65] This may be an indication that beons are such large clusters of lesser mesons, that pions inside are configuring into protons before they can escape the beon's interior.

When the right collection of the pions are trapped together within the beon, they may fuse together to become a proton. The stable protons then either stay in the interior of the beon until its surroundings decay away allowing the proton to escape, or meld with the mesons around it to form Δ or Λ baryons.

The charged beons, on occasion, also emit lepton particle-antiparticle pairs such as e^+-e^-, μ^+-μ^- and ν-$\bar{\nu}$. The Standard Model attributes this to the emission of a Z boson as shown in Fig. 3.13. However, this could happen because pions and kaons within the beon are decaying to either electrons or muons. The daughter particles are then trapped within the beon until some of its outer structure decays away. These decays are very rare (decay fraction $\Gamma_i/\Gamma > 10^{-5}$) but do occur.

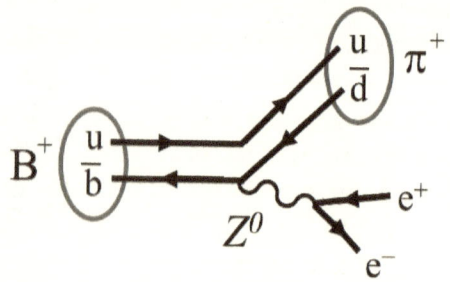

Fig. 3.13: Standard Model decay of B$^+$ to π^+, e$^+$ and e$^-$.
The $\bar{b}$ quark emits the neutral Z^0 boson, changing it into a $\bar{d}$ quark. That $\bar{d}$ pairs with the u quark from the original B$^+$ to form a π^+. The Z^0 boson decays into an electron – positron pair.

When all the dust settles, the charmed mesons D$^+$, D$^-$ and D^0 are made of kaons and pions, the charmed-strange mesons D$_s{}^+$ and D$_s{}^-$ are made of kaons and pions, and the bottom mesons B$^+$, B$^-$ and B^0 are made of D mesons, kaons and pions.

If the D mesons, both charmed and charmed-strange, and all the B mesons are all made of pions and kaons, then it appears the strange, charm and bottom quarks do not exist in mesons. The top quark supposedly decays into the bottom quark via W^- emission in the Standard Model. If there is no bottom quark, then there is nothing for the top quark to decay into, suggesting that it does not exist either.

Based on this sampling, it seems safe to assume that all the other mesons not considered here are ultimately made of pions and kaons. This means that it appears all the mesons are made of collections of electrons and positrons and not of the Standard Model quarks. Combined with the conclusions from the assessment of the baryons, it seems that the Standard Model quarks are not needed to describe the baryons and mesons.

If the proton is not made of quarks, then there appears to be no basis for believing that any of the baryons or meson are made of quarks. If the baryons are not made of quarks and gluons, and the mesons are not made of them, then it appears no particles are made of quarks and gluons and they do not exist at all.

The diagram in Fig. 3.14 shows the potential modifications required to the Standard Model particle matrix to accommodate the revised results of the electron-proton deep inelastic scattering. They showed that the proton is not made of quarks, and with no quarks in protons, it seems that quarks and gluons do not exist in baryons and mesons, at all. Without them, only the six leptons, three gauge bosons and the Higgs boson remain of the Standard Model, for now.

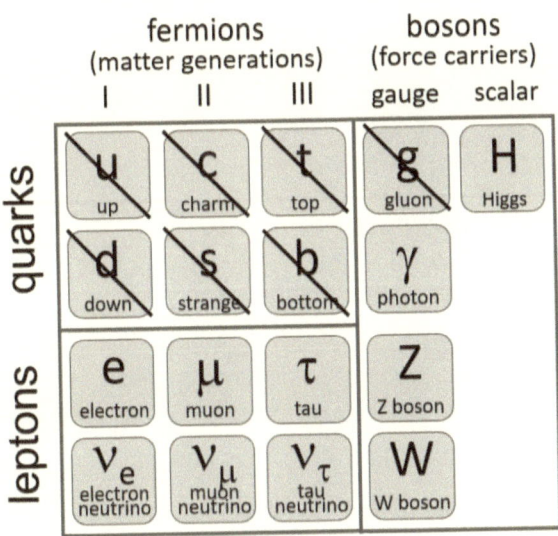

Fig. 3.14: Modified Standard Model
The Standard Model particle matrix showing the particles potentially eliminate by reanalyzing the electron-proton deep inelastic scattering. The up, down, strange, charm, top and bottom quarks are not needed to model the baryons or the mesons; and without them, the gluon is not needed to hold the quarks together.

The quark model of baryons and mesons is a wonderfully graphic and methodical way to characterize these particles and explain their behavior. However, as Gell-Mann pointed out from the start, the three particles with fractional charges is a "scheme" for making sense of these particles. It is a way to characterize them so that they can be arranged somewhat periodically.

But the deep inelastic scattering revealed that it is only a scheme, even though the results were misinterpreted to make the scheme appear to be real. The lack of quark observations in nature is a testament to the fact that they do not exist, regardless of the efforts to explain why the never appear.

4. What Comes Out of the Leptons

Leptons are considered fundamental particles in the Standard Model, meaning they are not thought to be made of smaller component particles. However, the scattering analyses in Chapter 2 proposes that the proton is made of eight bound pions, each containing about 229 electrons and positrons. Consequently, free pions appear to be made of 273 electrons and positrons.

The pion decays into a muon that appears to retain 207 of its electrons and positrons, seeming to challenge the notion that muons are fundamental particles. Unlike protons, for which probing particles have to be sent inside to determine their structure, leptons spew out an array of particles that reveal what they are made of.

4.1 Decay into an Electron and Neutrinos

When a free muon decays, it becomes a muon neutrino, an electron and an electron antineutrino,[66] or

$$\mu^- \to \nu_\mu + e^- + \bar{\nu}_e. \tag{4.1}$$

Per the Standard Model, the electron and electron antineutrino that appear as a result of muon decay come from the decay of a W^- boson. The W^- is a gauge boson in the Standard Model (see Fig. 1.1) responsible for carrying the weak force between particles, that has a mass of about 80 GeV. It supposedly forms with the muon neutrino when the muon decay occurs. The Feynman diagram in Fig. 4.1 shows the Standard Model muon decay.

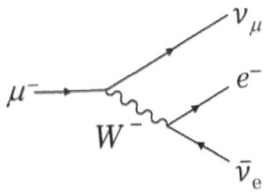

Fig. 4.1: Feynman diagram of muon decay.
The muon decays into a muon neutrino and a W^- boson. The W^- then decays into an electron and an electron antineutrino.

Similarly, when a tau particle decays, about 18% of the time it also decays into its neutrino (a tau neutrino), an electron and an electron antineutrino,[67]

$$\tau^- \rightarrow \nu_\tau + e^- + \bar{\nu}_e. \tag{4.2}$$

As with the muon decay, according to the Standard Model, the electron and electron antineutrino are the result of a W^- decaying. Fig. 4.2. shows the Feynman diagram of the tau decay in equation (4.2).

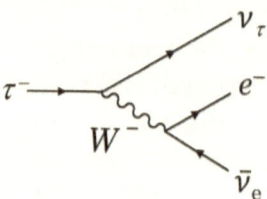

Fig. 4.2: Feynman diagram of tau decay.
The tau decays into a tau neutrino and a W^- boson. The W^- then decays into an electron and an electron antineutrino.

In both instances, the initial particle decays into its neutrino and a W^- boson, mildly suggesting that the difference between the particle and its neutrino is the W^- boson. However, the electron-proton deep inelastic scattering discussed in Chapter 2 indicates that the pion is made of particles the size of electrons and positrons, and Chapter 3 discusses how pions decay into muons. Knowing these, another hypothesis of how the muon decays can be formulated.

The μ^- has a unit negative charge. Therefore, if it is made of electrons and positrons, there must be an unpaired electron that acts as the muon's valence electron. Consequently, the muon must contain $2n + 1$ particles: $n + 1$ electrons and n positrons. With a mass equivalent to 206.768 electron masses, n for the free muon could be 102, but it is more likely 103. This gives it a total of 207 particles, 103 positrons and 104 electrons.

With this muon model, a scenario for μ^- decay can be devised from three observations. First, only one electron emerges from the μ^- decay. From that, it appears that during the decay, the muon's n positrons must annihilate n of its electrons. This would leave only the muon's lone unpaired valence electron and produce about 105 MeV of annihilation energy. Most of this energy is subsequently carried off by the decay particles.

Next, a muon neutrino appears when the μ⁻ decays. Since no matching antineutrino appears, it does not appear to be the product of a pair production during the decay. In fact, section 3.3 hypothesized that this neutrino is the product of pair production during the pion decay that formed the muon. It is a component of the muon before the decay. The annihilation of the muon's electrons and positrons during the decay sets it free. It is the central body that the electrons and positrons forming the μ⁻ assemble around.

Finally, an electron antineutrino appears along with the valence electron emitted during the μ⁻ decay. Electrons escaping the muon appear to become free electrons, but not without the creation of an electron antineutrino. It seems unlikely that an antineutrino would form without the formation of a neutrino as a result of pair production. The neutrino must somehow pair with the electron during the decay. This hints at the electrons inside the muon and proton – call them beta electrons (β^-) – being different from free electrons (e^-).

In the last chapter, it was speculated that muons are produced from charged pions, which are made of (beta) electrons and positrons that annihilate each other during its decay. The charged pions are 133 electron-positron pairs and an unpaired valence electron or positron. When the pions decay, their electrons and positrons randomly annihilate each other. The electrons and positrons continue the annihilations until there are only 103 pairs left, the number of electron-positron pairs in a muon. At that point, somehow the net charge of the partially decayed pion, along with its mass and energy from the annihilations produce a muon neutrino-antineutrino pair.

What is left of the pion, the 103 electron-positron pairs and either an unpaired electron or positron, captures the neutrino, if the unpair particle is an electron, or the antineutrino, if it is a positron. At this point, the pion has transformed into a muon or antimuon. The uncaptured neutrino or antineutrino becomes a decay product.

Once the muon forms, the beta electron-beta positron annihilations apparently continue, now causing the muon to decay. Similar to the pion, when all the beta electron-positron pairs of the muon have annihilated, it appears the charge of the final beta electron or beta positron plus its mass and the annihilation energy somehow produces an electron neutrino-antineutrino pair. The lone beta electron captures the neutrino to become a free electron or the beta positron captures the antineutrino to become a free positron.

This makes the muon a composite particle made of its neutrino surrounded by beta electrons and beta positrons. The electron also appears to be composite, consisting of a beta electron coupled to an electron neutrino. Since tau particles decay into muons and electrons, they, too, appear to be composite. Only beta positron, beta electron and the three neutrinos now appear to be fundamental.

4.2 The Structure of Free Electrons

During muon decay, it appears a beta electron produces a neutrino-antineutrino pair using annihilation energy, from which it captures the neutrino to become a free electron. This conversion process does not appear to be limited to muon decay. Beta electrons seem to produce neutrino-antineutrino pairs to transform into free electrons whenever beta decays occur.

Radioactive nuclei appear to be ones that have an excess of particles. An excess of alpha particles causes alpha decay, an excess of electrons or positrons, beta decay, and an excess of electron-positron pairs causes gamma emission. In a nucleus containing an excess beta electron or positron, it seems the excess beta electron or beta positron initiates the decay by igniting the annihilation of some beta electron-positron pairs. This creates a pool of annihilation energy within the nucleus. The annihilation provides the excess charged particle with the energy it needs to spawn a neutrino-antineutrino pair.

If the excess particle is a beta electron, β^-, it captures the neutrino to become a free electron, e^-, and both it and the leftover antineutrino exit the nucleus. If the excess particle is a beta positron, β^+, it captures the antineutrino to become a free positron, e^+, and it and the neutrino exit the nucleus.

The reaction sequences below show examples of how these negative (^3H) and positive (^{13}N) beta decays likely occur. They show the internal annihilations occurring creating energy (ε) and an altered form of the isotope (X^*). The energy engulfs an excess β^- or β^+ inside the ( ) spawning a neutrino-antineutrino pair inside the { }.

If the excess beta is a β^-, it couples with the neutrino inside { }, becoming a free electron, e^-; which, along with the residual antineutrino, exit the nucleus (^3He). If the excess beta is a β^+, it couples with the antineutrino inside { } and a free positron and neutrino exit the nucleus (^{13}C). Everything inside the braces [ ] below happens inside the original nucleus. The ε^* is the energy after the pair production.

$$^3_1H \rightarrow {}^3_1H^*[\varepsilon]$$
$$\rightarrow {}^3_2He^*[\varepsilon(\beta^-)]$$
$$\rightarrow {}^3_2He^*[\varepsilon^*(\beta^- + \{v_e + \overline{v}_e\})]$$
$$\rightarrow {}^3_2He^*[\varepsilon^*(\{\beta^- + v_e\} + \overline{v}_e)]$$
$$\rightarrow {}^3_2He + e^- + \overline{v}_e + \varepsilon^*,$$

$$^{13}_7N \rightarrow {}^{13}_7N^*[\varepsilon]$$
$$\rightarrow {}^{13}_6C^*[\varepsilon(\beta^+)]$$
$$\rightarrow {}^{13}_6C^*[\varepsilon^*(\beta^+ + \{\overline{v}_e + v_e\})]$$
$$\rightarrow {}^{13}_6C^*[\varepsilon^*(\{\beta^+ + \overline{v}_e\} + v_e)]$$
$$\rightarrow {}^{13}_6C + e^+ + v_e + \varepsilon^*.$$

The electron capture process appears to support this model of the electron. There, a free electron, e^-, is drawn into the nucleus, reducing its charge and mass. At some point during the capture process, an electron neutrino appears. Based on the model offered above, that neutrino was part of the free electron.

Electron capture is equivalent to β^+ decay, which reduces the mass and charge of the nucleus by emitting a positron from the nucleus. Therefore, upon entering the nucleus, the newly acquired electron must annihilate a beta positron. To do this, the β^- part of e^- likely separates from the neutrino. This is probably why the neutrino appears during the process.

4.3 Leptons and Their Neutrinos

The diagram in Fig. 4.3 shows the muon decay with the muon neutrino emerging along with a β^{*-} instead of a W^-. The β^{*-} is the sole surviving β^- of the mass annihilation of the beta electrons and beta positrons that made up the bulk of the muon. The asterisk (*) indicates that the β^- is immersed in a pool of annihilation energy. The β^{*-} essentially "decays" into a free electron and an electron antineutrino by creating a neutrino-antineutrino pair from the annihilation energy produced during the decay and capturing the neutrino.

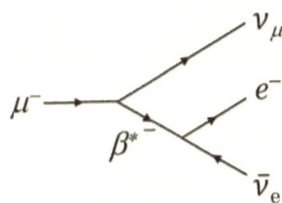

Fig. 4.3: **Feynman diagram of the modified muon decay**
Diagram showing the muon neutrino and the valence beta electron emerging from the initial annihilation of the 103 beta electron-beta positron pairs. The beta electron essentially decays into a free electron and an electron antineutrino.

One might propose that the muon neutrino that appears during the decay could also be the product of pair production like the electron neutrinos. There, only the electron antineutrino is visible (indirectly). However, even though the electron neutrino is not visible, it is accounted for as part of the free electron. The muon neutrino is there, but where is the muon antineutrino after the decay?

The answer is – there is not one. As with the electron in muon decay, when the muon is the product of a decay, like pion decay (section 3.3), a muon antineutrino also appears. In fact, when that happens in one branch of tau decay, one also appears.

When a τ^- decays, in addition to the decay branch shown in equation (4.2), about 17% of the time it can also decay into a tau neutrino, a muon and a muon antineutrino,[68]

$$\tau^- \rightarrow \nu_\tau + \mu^- + \bar{\nu}_\mu. \tag{4.3}$$

Here, when the free muon forms, a muon antineutrino also appears like the electron antineutrino appearing when the free electron forms during muon decay. This signals that a muon neutrino-antineutrino pair formed during the tau decay, from which the muon neutrino was captured to form a free muon.

The remaining muon antineutrino is left as a decay product of the tau decay like the electron antineutrino is after free electron formation during muon decay. The Feynman diagrams of the τ^- decays in equations (4.2) and (4.3) are shown in Fig. 4.4. The X in the diagram on the left represents a collection (> 207) of β^- and β^+ particles.

But, in a decay event starting with a muon like in Fig. 4.3, the muon already exists. Therefore, the neutrino appearing during the decay does not come from a pair production ignited to form a particle. Consequently, less any other viable source, the muon neutrino that appears during muon decay must come from within the muon.

The tau parallel to the muon model is a τ^- made of $2n + 1$ particles: $n + 1$ electrons and n positrons, but with a mass of 3,477.15 electron masses, n must be about 1,738 for the τ^-. Like the muon, the electrons and positrons making up the τ^- apparently orbit a tau neutrino. The τ^- decay should occur in the same manner as the decay of the μ^-; and indeed, the branch of its decay shown in Fig. 4.2 does appear to do so. However, because of its large mass, τ^- can decay along other paths, including forming the less massive muon.

About 5.5% of the time, charged D$_s$ mesons decay into tau leptons forming tau neutrino-antineutrino pair,[69] just as pions form muon neutrino-antineutrino pairs when they decay into muons. Consequently, when the tau decays a tau neutrino appears, just as a muon neutrino appears in Fig. 4.3 when the muon decays, and an electron neutrino appears when the electron disintegrates during electron capture. All three particles are leptons. The appearance of their respective neutrinos when they disintegrate seems to show that the core of a lepton is its neutrino.

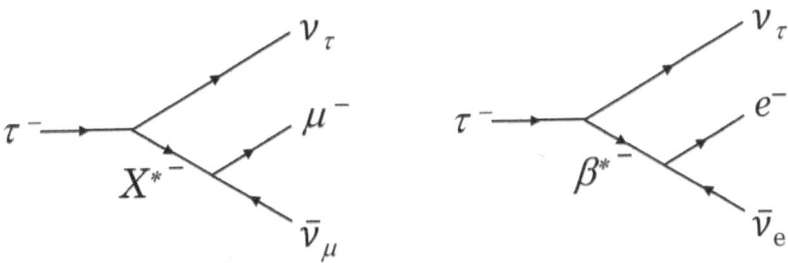

Fig. 4.4: Feynman diagrams of modes of tau decay.
In both, the tau decays into a tau neutrino and another lepton and its antineutrino. The X*⁻ in the diagram on the left is an unknown precursor to the muon that forms.

The Feynman diagram in Fig. 4.5 shows how electron capture can be interpreted as the decay of the electron like the decay of its cousins the muon and the tau. As the electron gets pulled into the nucleus, it splits (decays) into an electron neutrino and a beta electron. Its core neutrino is freed as is its valence beta electron, just as is the case in muon decay and in one version of tau decay.

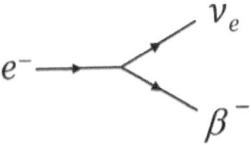

Fig. 4.5: Pseudo-decay of the electron during electron capture.
Feynman diagram of the free electron as it is pulled into the nucleus. It splits into an electron neutrino and a beta electron.

If the electron is a beta electron coupled with an electron neutrino and the muon about 207 beta electrons and positrons orbiting a muon neutrino, then is the tau, at about 3,477.15 electron masses, about 3,477 beta electrons and positrons orbiting its tau neutrino?

4.4 A Model of the Electron

The lepton decays discussed in Section 4.3 suggest that the electron is made of a beta electron coupled to an electron neutrino. It can be shown that the electron magnetic moment supports an electron model with the beta electron orbiting the neutrino.

At first glance, it seems the electron magnetic moment, μ_e, should be equal to the Bohr magneton, μ_B, but it is not. The magnitude of the electron's magnetic moment, $\mu_e = -9.28476 \times 10^{-24}$ J/T, is slightly greater than the Bohr magneton, $\mu_B = -9.27401 \times 10^{-24}$ J/T. The actual magnetic moment is 1.0011659 times greater than the Bohr magneton or $\mu_e = 1.0011659\ \mu_B$.

In the Standard Model, this difference is attributed to quantum loop effects that cause the gyromagnetic ratio of a particle, g_p, to deviate from the Dirac equation value of $g = 2$. The loop effects are characterized by the anomalous magnetic moment of the particle, a_p, where $a_p = (g_p - 2)/2$. For the electron, the anomalous magnetic moment is $a_e = (\mu_e/\mu_B) - 1$, or $a_e = 0.0011596$.[70] This makes the gyromagnetic ratio of the electron $g_e = 2.0023192$. However, the difference also provides a clue to the structure of the electron.

The dimensions of the magnetic moment, Joules per Tesla (J/T), reduce to Coulombs meters-squared per second (C-m²/s). In the macroscopic world, this is usually interpreted as the product of a current (C/s) and an area (m²). The magnetic moment is the product of a current of electricity moving through a loop of wire, times the area of the loop. This interpretation produces the Bohr magneton as the magnetic moment of the electron.

However, the units of magnetic moment can also be interpreted as a "moment of charge" (C-m²) – similar to the moment of inertia for mass (kg-m²) – times a frequency of revolution (s⁻¹). This interpretation seems more appropriate for determining the moment of a single particle versus the many particles flowing through a wire. Now, the magnetic moment of a charged particle becomes

$$\mu = I_q \nu, \tag{4.4}$$

where I_q is its moment of charge and ν is its spin frequency.

The moment of inertia, I_m, (mass) for a solid sphere is $I_m = \frac{2}{5} mr^2$, where m is the mass of the sphere and r is its radius. Replacing the mass in the expression with charge makes the moment of charge, I_q, for a solid sphere $I_q = \frac{2}{5} qr^2$, where q is the charge of the particle. The moment of inertia for a thin-shelled hollow sphere is $I_m = \frac{2}{3} mr^2$, making the moment of charge for the hollow sphere $I_q = \frac{2}{3} qr^2$.

The classical radius of the electron is 2.8×10^{-15} m, roughly three times the proton radius of 0.9×10^{-15} m, but its mass is 1,836 times smaller than the proton's mass. Assuming the density of the matter forming the electron is the same as that making up the proton, then for the electron to be larger, it must not be solid. This can be the case if the mass of the electron is concentrated in a small particle that conserves the density of nuclear matter, in an orbit having the electron's classical radius.

The model proposed here is a composite electron with a beta electron in a high-frequency orbit around a neutrino. If the orbit of the beta electron precesses about an axis through the neutrino, the beta electron creates a virtual thin-shelled hollow sphere around the neutrino. This is consistent with the size difference between the electron and proton. This configuration would make the electron's moment of charge $I_e = \frac{2}{3} q_e r_e^2$; and its magnetic moment, $\mu_e = \frac{2}{3} q_e r_e^2 v$. A diagram of this electron model is shown in Fig. 4.6.

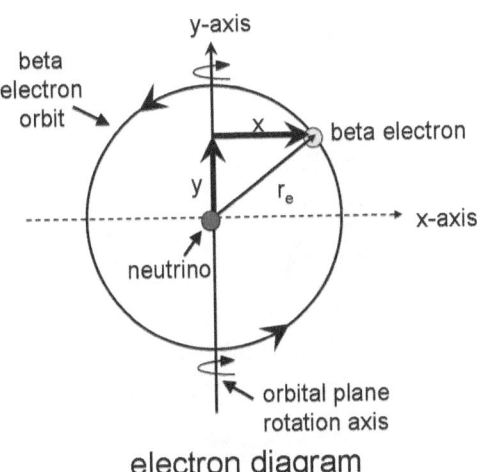

electron diagram

Fig. 4.6: Diagram of the electron model
The electron depicted as a beta electron orbiting a neutrino. As the beta electron orbits, the orbit precesses about the y-axis creating the illusion of a hollow sphere.

While the high-frequency, precessing orbit of the beta electron around the neutrino approximates a thin-shelled hollow sphere; in reality, it is a beta electron in a high-frequency orbit around a neutrino. It turns out that the moment of inertia for a particle whose motion approximates a thin-shelled hollow sphere is slightly greater than the moment of inertia of a hollow sphere. In addition to the moment of inertia of the hollow sphere, the moment of inertia of the particle must also be accounted for using the parallel axis theorem.[71]

Briefly, the parallel axis theorem states that the moment of inertia of a body with respect to an axis not through the body, I', equals the moment of inertia of the body, plus the product of its mass and the average distance squared it is from the desired axis. That is, $I' = I + md^2$, where d^2 is the average distance squared from the axis.

Paralleling this concept to the moment of charge gives $I' = I + qd^2$, or for the electron, $I_e = I_{be} + q_{be} d^2$. Here, the subscript, "*be*", denotes the beta electron and d^2 is the average square distance the beta particle is from the rotation axis during one complete orbit.

Assuming the beta electron is a solid sphere, its moment of charge is $I_{be} = 2/5\, q_{be}\, r_{be}^2$. The average distance squared that the beta particle is from the moment axis during its orbit is determined by assuming it follows a circular orbit. The equation of the orbit, if it is in an x − y plane, is $x^2 + y^2 = r_e^2$, or $x^2 = r_e^2 - y^2$. This relationship makes the average distance squared

$$d^2 = \frac{\int_0^{r_e} x^2 dy}{\int_0^{r_e} dy} = \frac{\int_0^{r_e}(r_e^2 - y^2)dy}{\int_0^{r_e} dy} = \frac{r_e^3 - \frac{1}{3}r_e^3}{r_e} = \tfrac{2}{3}r_e^2. \quad (4.5)$$

Therefore, the product of the beta electron charge and its average distance squared from the moment axis during one complete orbit is

$$q_{be}\, d^2 = {}^{2}/_{3}\, q_{be}\, r_e^2 = {}^{2}/_{3}\, q_e\, r_e^2,$$

since $q_{be} = q_e$. This is equal to the hollow-sphere electron moment of charge. Using it and the beta electron moment of charge, the actual moment of charge for the electron, $I_e = I_{be} + q_{be} d^2$, becomes

$$I_e = {}^{2}/_{5}\, q_{be}\, r_{be}^2 + {}^{2}/_{3}\, q_e\, r_e^2,$$

or

$$I_e = (\tfrac{3}{5}\frac{r_{be}^2}{r_e^2}+1)(\tfrac{2}{3}q_e r_e^2). \quad (4.6)$$

If the hollow sphere moment of charge corresponds to that of the Bohr magneton, then the expression in the first set of parentheses in Eq. (4.6) appears to correct it to the actual electron moment of charge. Assuming the frequency used to calculate the magnetic moment is the same for both the electron and the Bohr magneton, the expression in the first set of parentheses is the 1.0011596 factor that corrects the Bohr magneton to the actual electron magnetic moment. This makes the anomalous magnetic moment of the electron

$$a_e = \tfrac{3}{5} \frac{r_{be}^2}{r_e^2},\tag{4.7}$$

or

$$0.0011596 = \tfrac{3}{5} \frac{r_{be}^2}{r_e^2},\tag{4.8}$$

which makes

$$r_e = 22.75\, r_{be}.\tag{4.9}$$

This says that the radius of the free electron, which is a composite made of the beta electron and a neutrino, is about 23 times the radius of the beta electrons and beta positrons that reside within the proton. It also suggests that the *g*-factor that adjusts the Bohr magneton to the electron magnetic moment is due to the presence of the small beta electron in the free electron.

4.5 A Model of the Muon

Muon decay indicates that, like the electron, the muon has one of its neutrinos as a component. Its mass of 206.768 free electron masses and unit negative charge suggests a muon model containing 207 particles, 104 beta electrons and 103 beta positrons. Assuming they configure themselves in orbits around the muon neutrino at the center of the muon, the muon takes the form of a large electron. It is an electron with 207 beta particles orbiting its neutrino instead of one.

As with the electron, the magnetic moment of the muon can give some insight into how the beta electrons and beta positrons configure themselves within the muon. Since the muon and the electron have the same charge, the muon equivalent of the Bohr magneton is the Bohr magneton equation with the muon mass replacing the electron mass. Therefore, the "muon" magneton, μ_M, is the Bohr magneton times m_e / m_μ.

The muon magnetic moment is 1.0011659μ_M, making its anomalous magnetic moment $a_\mu = 0.0011659$ compared to $a_e = 0.0011596$ for the electron. Having an anomalous magnetic moment nearly identical to that of the electron is a strong indication that the beta particles inside the muon are in orbits similar to that of the single beta electron in the free electron in Fig. 4.6. By replacing 0.0011596 in equation (4.8) with 0.0011659, the muon equivalent to equation (4.9) becomes

$$r_\mu = 22.65 r_{be}. \tag{4.10}$$

All the beta electrons and beta positrons must be orbiting the central muon neutrino at that radius, and their orbital planes must rotate about an axis similar to Fig. 4.6. This would produce a much denser version of the free electron. A particle that is essentially the same size as a free electron, but about 207 times more massive.

If the muon is made of 207 beta particles, then the simplest model of it is a sphere containing 207 beta particles, each scribing out an orbit like the beta particle orbit in the electron. This would form 207 meridians around the central neutrino.

The problem with this model is that each beta particle orbit takes up two spaces on the muon equator because it passes through the equator twice. Therefore, 207 orbiting particles, side-by-side along the equator of the muon makes the circumference of the muon equator 414 beta particle diameters, or 828 beta particle radii. Since the circumference of the muon is 2π times its radius, the relationship between the beta electron radius and the muon radius becomes

$$2\pi r_\mu = 816 r_{be},$$

$$r_\mu = 129.87 r_{be}. \tag{4.11}$$

This muon radius is much greater than the muon radius required to produce the anomalous magnetic moment of 0.0011695 determined in equation (4.10), which means the model is not right.

One way to modify the muon model to align better with the electron is to put six beta particles, three beta electrons and three beta positrons, in each orbital around the muon neutrino. This would reduce the number of orbitals around the neutrino by a factor of six, from 207, to 34, with three betas left over. It now makes the circumference of the muon equator 136 beta-electron radii, instead of 828.

With this model

$$2\pi r_\mu = 136 r_{be}, \text{ or } r_\mu = 21.65 r_{be}, \qquad (4.12)$$

much closer to equation (4.10). Realizing that equation (4.10) represents a 207-electron muon, one more orbital, which contains only three beta particles, must be added to the model. It must contain either two electrons and one positron or two positrons and one electron. This adds two beta electron diameters or four beta electron radii to the circumference of the muon equator. It makes equation (4.12)

$$2\pi r_\mu = 140 r_{be}, \text{ or } r_\mu = 22.28 r_{be}, \qquad (4.13)$$

even closer to equation (4.10).

Fig. 4.7 shows what the muon might look like. The hemisphere shown in the figure contains about half of the beta electrons and beta positrons making up the muon. The other half is on the hidden hemisphere on the other side of the muon. The beta particles orbiting the central muon neutrino are all moving from bottom to top in the hemisphere shown, and top to bottom in the hidden hemisphere.

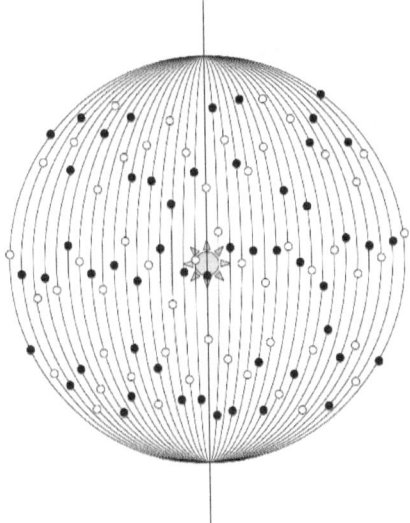

Fig. 4.7: Diagram of the muon model
The muon depicted as 207 beta electrons (black dots) and positrons (white dots) orbiting a muon neutrino (star) in 35 orbits. As the beta particles orbit, the orbits precess about the axis creating the illusion of a hollow sphere.

Alternating beta electrons (black dots) and beta positrons (white dots) orbit a muon neutrino (star) to form the muon. The figure shows 53 of the 104 beta electrons and 52 of the 103 beta positrons in orbit around the muon neutrino, the gray star at the center. The diagram also shows the axis of rotation of the muon, about which, the orbital planes of the beta particles in a free muon precess.

This model of the muon shows why muons and electrons are sometimes interchangeable in physical situations. A muon is truly just a heavy unstable electron. Beta particles inside the muon orbiting the muon neutrino explain why the free electrons do not retain their electron neutrinos while in the proton. They no longer need them because they have the muon neutrino to orbit. This also is why the free electrons do not retain their electron neutrinos upon entering the nucleus during electron capture.

From equations (4.9) and (4.10), the radius of the electron is essentially the same as the radius of the muon. Section 2.6 showed that if the charge is evenly distributed throughout the proton, then the radius of the pions in it is about 0.33×10^{-15}m. In the next chapter, it is shown that when pions decay into muons, the muons are the same size (radius) as pions. Therefore, the radius of the electron would also be the same as that of the pion, or about 0.33×10^{-15}m.

An electron radius of 0.33×10^{-15}m means that the proton radius appears to be about 2.65 times the radius of the electron. This electron radius is nearly an order of magnitude smaller than the classical electron radius of 2.8×10^{-15}m.

4.6 A Model of the Tau

The Standard Model predicts the anomalous magnetic moment of the tau (0.0011772) to be slightly greater than the anomalous magnetic moment of the muon (0.0011695), which is slightly greater than that of the electron (0.0011596). This suggest that, in the Standard Model, the tau is just an extension of the muon and electron.

If the tau is just a heavy muon, then one might expect it to be similar in structure to the muon. This is not an unreasonable expectation since it can decay into a muon. Given the distinctive structure of the muon, the tau likely shares the major features of it. Consequently, one model of the tau could be a tau neutrino inside a shell of 3,477 beta particles, 1,738 beta electron pairs and a valence beta electron or positron.

If the tau retains the 35 orbitals of the muon, each orbital would contain about 100 beta particles. Since equation (4.13) puts the muon radius at about 22 times that of a beta electron, a muon orbital can hold a maximum of about 70 beta particles. To hold 100 beta particles, the circumference of the orbital must be increased by about 43% to 31.8 beta electron radii. However, at this radius, the betas would be crammed into the tau; so, for this model, it is likely even greater.

This model could facilitate the decay of a tau into a muon or electron. For a muon, it need only burn off the appropriate number of electron-positron pairs (1,635) through annihilation and produce a muon neutrino-antineutrino pair in the process. For the electron, all the beta electron-positron pairs must get annihilated and an electron neutrino-antineutrino pair created.

Because a specific number of beta electron-positron pairs must be annihilated to form the muon, it seems more likely that a second shell of beta pairs surround the shell containing the 207 beta particles that will form the muon. That shell contains the 1,635 pairs that must annihilate to form the muon. Now, the particle starts out as a tau, then quickly burns off its outer shell, becoming a muon.

It turns out that, because of its short lifetime, the tau's magnetic moment cannot be determined by the conventional means used to measure the magnetic moments of the muon and the electron. Consequently, no experimental value of its anomalous magnetic moment, a_τ, has been measured, to date. However, a number of indirect techniques have been employed in an attempt to place limits on the value of a_τ. The latest range is $0.013 < a_\tau < -0.056$.[72]

In most cases, the limits indicate that the actual anomalous magnetic moment of the tau is significantly greater than the Standard Model prediction. This suggests that the Standard Model does not provide a good model of the tau. It also suggests that, contrary to the Standard Model implication, the tau is not an extrapolation of the muon and electron. Its structure is far more complex, and its magnetic moment is not determined by the same parameters.

The uncertain anomalous magnetic moment could support the shell-in-shell model of the tau mentioned earlier. The outer shell would have a radius greater than the inner shell radius, which is apparently what produces the anomalous magnetic moment predicted by the Standard Model. The larger radius of the outer shell would cause the tau to have a greater anomalous magnetic moment.

A review of the tau decay modes reveals that the tau can decay into either multiple mesons (π's and K's) or lepton pairs (μ^+-ν_μ, μ^--$\bar{\nu}_\mu$, e^+-ν_e or e^--$\bar{\nu}_e$), but not both simultaneously. This suggests that the particles being called taus may actually be a mixture of two different particles. One of the particles is like the leptons e and μ, call it τ_L, and the other is like the mesons K and D, τ_M. In fact, D_s mesons decay into tau leptons about 5.5% of the time, and even B mesons can decay into taus, although rarely (< 0.06%). The lifetime of the D, 5.0×10^{-13} seconds, is also suspiciously close to that of the τ, 2.9×10^{-13} seconds.

Particles thought to be taus decay into lepton pairs about 37% of the time, probably the τ_L's and into mesons the remaining 63%, the τ_M's. It may be that the ones that decay into leptons are real taus and the others, a meson whose mass is close to that of the tau. In many instances, the τ^- is thought to decay into a tau neutrino and a collection of as many as seven pions,[73] some charged and some uncharged.

Of course, the tau neutrinos cannot be seen by detectors and only are implied through missing mass. Likewise, particles like neutral pions are also invisible to the detectors. In instances where they are also part of the decay, the neutral pions may be carrying the missing mass attributed to the tau neutrino. There may be no tau neutrino involved in those decays at all.

Given its mass, the tau could be a collection of pions, similar to the proton. With 3,477 electrons and positrons, the tau could contain as many as 12 pions, one and one half as many as the proton. As discussed in Chapter 3, charged pions have a mass of 139.6 MeV. If they are made of just beta electrons and beta positrons, they contain about 273 betas – 136 positrons and 137 electrons. The mass of a neutral pion is 135 MeV or about 264 electron masses, making a model of it 132 electrons and 132 positrons.

A τ^- made of 3,477 particles could be made of 12 charged pions – six positive and six negative – with 201 electrons and positrons left over. Some of the pions in the tau model could be neutral, making it likely that the model contains 13 or more pions. The 12 pions allow for virtually no mass defect to produce binding. An extra pion could produce about 15 - 20 MeV/particle binding energy.

5. Mesons and Protons, Again

The final model of the muon in section 4.5 has 207 beta electrons and beta positrons in 35 orbits around a muon neutrino. Charged pions decay into muons. Therefore, this is the configuration that remains after a charged pion decays into a muon. From the muon model, a likely model of the charged pion and description of its decay process can be hypothesized.

5.1 Pion Model Revisited

The charged pion appears to be made of 273 beta particles – 136 beta electron-positron pairs and an unpaired valence beta particle. The free muon is made of 207 beta particles – 103 beta electron-positron pairs and a valence beta particle. Therefore, during the charged pion decay, 33 of the 136 pairs of beta electrons and beta positrons making up the pion must annihilate to leave the 103 pairs and an unpaired beta electron or beta positron that make up the muon. The beta particles remaining after 33 pairs annihilate must also arrange themselves into the muon configuration of 207 beta particles in 35 orbits of six beta particles orbiting about the center of the particle.

One way for this to always happen during pion decay, given the muon configuration, is to have 33 beta electron-positron pairs and the unpaired beta electron or positron clustered in the center of the pion, where the muon neutrino is in Fig. 4.7. The remaining 103 beta electron–beta positron pairs are particles in orbit around this central cluster of beta particles, like those orbiting the muon neutrino in the muon model. A diagram of this configuration is shown in Fig. 5.1.

Now, when the beta particles in the center of the pion all annihilate each other, the shell of the muon remains. The annihilations leave a pool of energy and the unpaired beta immersed in the energy at the center of the decaying pion. The annihilations in the pion always stop at 33 pairs because that is all there are to annihilate. The beta particles are in 35 orbits about the center of the decaying pion because that is the configuration they started in as part of the pion. The muon inherits its beta particle shell configuration from the pion.

61

Once the 33 annihilations have occurred (and they probably happen quickly), the unpaired beta particle spawns a muon neutrino – antineutrino pair from the annihilation energy. If the pion was negative, the neutrino takes up residence at the center of the decaying pion. If it was positive, the antineutrino stays at the center.

One of the 35 orbitals has only two beta particles in it, a beta electron-positron pair. The unpaired beta particle joins the pair in that orbital. This gives that orbital a charge and the potential to initiate the cascade of annihilations that ultimately decays away the muon.

The other neutrino escapes as the other half of the lepton pair, turning the decaying pion into a muon. The diagram in Fig. 5.1 shows what the pion might look like. A muon with a cluster of beta electrons and beta positrons at its center instead of a muon neutrino.

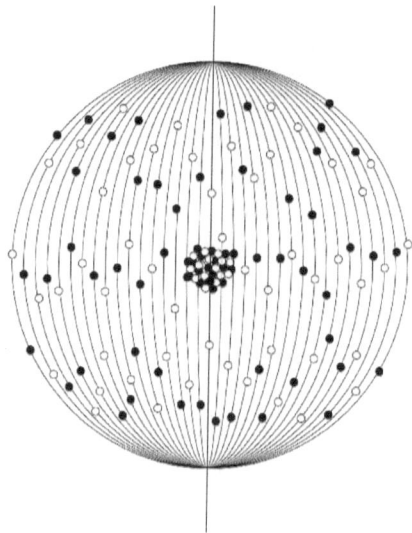

Fig. 5.1: Diagram of the pion model
The pion depicted as 103 beta electrons (black dots) and 103 beta positrons (white dots) orbiting a cluster of 33 beta electron-positron pairs and an unpaired beta particle (center) in 35 orbits. As the beta particles orbit, the orbits precess about the axis creating a virtual hollow sphere. The lines show the meridians of the orbits.

This pion model has at least three advantages to it over the Standard Model pion. First, it is made of particles readily and extensively found in nature. Secondly, its structure provides it an easy path to becoming the muon it is observed to decay into. And finally, the processes that cause its decay – pair annihilation and pair production – are also commonly seen occurring in nature.

The neutral pion, with a mass of 134.977 MeV, is 264 free electron masses, nine less than the charged pions. Since it is neutral, that means it is made of 132 beta electron-positron pairs. It decays into two photons 99% of the time. A neutral pion configured essentially the same as the charged pions can facilitate the observed decay.

If the neutral pion is a shell of 103 beta electron-positron pairs orbiting a central cluster of 29 beta electron-positron pairs, there are two collections of beta electron-positron pairs capable of complete annihilation. Without a valence beta particle, the central cluster of beta particle pairs can annihilate into a photon of energy. Once it is gone, the beta particle pairs in the orbital shell can annihilate each other, producing the second photon.

The neutral pion should look much like the charged pions shown in Fig. 5.1. The only difference is that instead of 33 beta electron-positron pairs and a valence beta particle in the central core, the neutral pion just has 29 beta electron-positron pairs in its central core.

The other 1% of neutral pion decays are into a free electron, a free positron and a photon. The neutral pion model offered here can also accommodate this decay.

During annihilation of the pion core, a beta electron or beta positron could spawn a neutrino-antineutrino pair out of the annihilation energy before all the electrons and positrons have annihilated. Then, one of the beta electrons remaining in the cluster could capture the neutrino to become a free electron and one of the beta positrons could capture the antineutrino to become a free positron.

The two new particles carry away the annihilation energy, so no photon forms from the annihilation of the pion core. However, the beta electron-positron pairs in the pion shell still also annihilate, producing a photon. Now, what was a neutral pion has become a free electron, a free positron and a photon. Although this rarely happens during neutral pion decay, this is how it can happen.

5.2 Another Look at the Kaon

The pion models above can be extended to create a model of the kaon. Apparently, at least two configurations of the kaon exist: one that decays into a collection of pions, call it K_π, and another that decays into a muon and a muon antineutrino, call it K_μ. The one that becomes a muon, K_μ, is likely similar to the pion, but with more beta electron-positron pairs at its center.

From section 3.4, the charged kaon is likely made of 967 beta electrons and beta positrons, 483 beta electron – beta positron pairs and one unpaired beta electron or positron. About 63% of all charged kaons decay into muons and muon antineutrinos, so they are K_μ's.

Assuming that, as with the pion, 103 pairs of beta electrons and positrons are in orbits about the center of this kaon, 380 pairs of beta electrons and beta positrons plus the unpaired beta electron or positron must be clustered at the center of the kaon. It might look something like the diagram in Fig. 5.2. This type of kaon is just a heavier version of the pion. Now, like with the pion, the central cluster of beta particle pairs will annihilate.

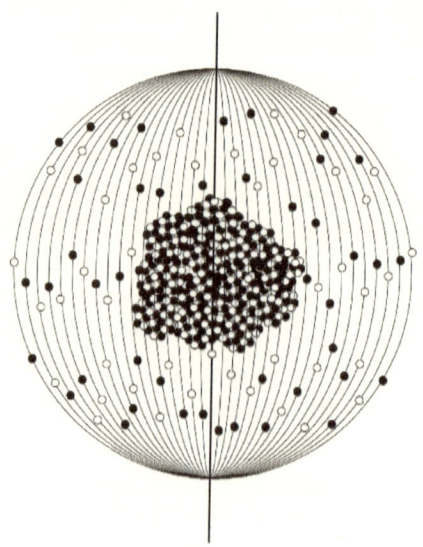

Fig. 5.2: Diagram of the K_μ kaon model
The K_μ kaon depicted as 103 beta electrons (black dots) and 103 beta positrons (white dots) orbiting a cluster of 380 beta electron-positron pairs and an unpaired beta particle (center) in 35 orbits. As the beta particles orbit, the orbits precess about the axis creating a virtual hollow sphere. The lines show the meridians of the orbits.

Once they have all annihilated, the unpaired beta particle is left in a large pool of annihilation energy, as was the case in pion decay. Again, the beta particle spawns a muon neutrino – antineutrino pair. If the unpaired beta is an electron, the neutrino stays at the center of the decaying K_μ kaon, making it a negative muon. If it is a positron, the antineutrino takes the center, turning the kaon into a positive muon. The other neutrino exits the configuration.

As with the pion, the unpaired beta joins the other betas in orbit about the central neutrino. The unpaired beta goes to the orbital containing only one beta electron-positron pair, making the particle a muon. This leaves the orbital unbalanced and eventually initiates the annihilations that cause the muon to decay into an electron.

In both the pion and the K_μ kaon models, beta electrons and beta positrons cluster at the center of the particle. The pion has 67 beta particles at its center, while the K_μ kaon has 761 beta particles. However, both particles have 206 beta particles orbiting the central cluster. How is it that a cluster of 67 beta particles would attract the same number of beta particles in its orbitals that a cluster of 761 would attract? There may be a simple reason for this.

Even though the K_μ kaon has over ten times the number of central beta particles as the pion; both particles have only one valence particle that gives it either a unit positive or unit negative charge. It may be that the unit charge of the central cluster is what determines the configuration of the shell of orbitals. This sounds plausible, except there is the case of the muon. The muon appears to have the same beta shell configuration of the pion and the K_μ kaon, but its model has no charge at its center. This observation segues into the most likely reason for the similarities. All the muons were once pions or K_μ kaons.

The particles being discussed here are not enduring particles. They come and go in fractions of a second. Consequently, it is not unreasonable to assume that anytime a muon appears, at some time in the past, it was either a K_μ kaon that decay into the muon, or it was once a pion.

Therefore, the beta particle shell that the muon has is inherited from the K_μ kaon or the pion. The muon had nothing to do with the formation of the shell it possesses. The shell was created to accommodate the charged central clusters of the K_μ kaon and the pion and handed down to the muon during their decays.

This means that the valence particles of the clusters of beta particles appear to have determined the number of orbitals, the number of particles in the orbitals and probably the radius of the orbitals. Once the orbitals were established, when the centers of the K_μ kaon or the pion annihilated away, their orbital shells were somehow self-sustaining, and remained intact for their muon descendants.

About 28% of the charged kaons decay into two or three pions. These are the K_π kaons. That multiple pions are coming out of this kaon indicates that the pions are components making up this kaon. Although 20% of the kaons decay into only two pions while about 8% decay into three, a few rare K^+ decay modes reveal that the K_π is probably made of four pions. There are several kinds of K_π kaons.[74]

Some K^+ particles decay into three π^0 particles, a positron and an electron neutrino ($K^+ \to \pi^0 \pi^0 \pi^0 e^+ \nu_e$). Here, assuming the e^+ - ν_e pair results from the decay of a π^+ particle, the K^+ appears to have been made of three π^0 particles and a π^+ particle. Other K^+ particles decay into two π^+ particles, a π^- particle and a γ particle ($K^+ \to \pi^+ \pi^+ \pi^- \gamma$). The γ here results from the decay of a π^0 particle, making the kaon two π^+ particles, a π^- particle and a π^0 particle. Another K^+ particle decays into a π^+ particle and three γ particles ($K^+ \to \pi^+ \gamma \gamma \gamma$). Here, like the first example, the K^+ appears to have been made of a π^+ particle and three π^0 particles.

These three examples indicate that the K_π meson is likely made of four pions. When it decays into only two or three particles, one or two of the π^0 pions forming it have decayed and distributed their energy among the remaining particles. About one-third of the K^0_L neutral kaons decay into either three π^0 particles (19.5%) or a π^+, a π^-, and a π^0 (12.5%). This makes them appear to be K_π kaons made of four π^0 particles, and a π^+, a π^-, and two π^0 particles, respectively.

Because their structures are so different, one wonders if the mass of the K_π kaons is the same as that of the K_μ kaons. The mass of four pions is significantly higher than the quoted mass of a kaon and might be an indication that the mass of the K_π kaon is greater than the K_μ kaon mass. However, for now, they are assumed the same.

The charged kaons have a mass of 493.677 MeV or about 967 free electron masses. The mass of three neutral pions and one charged pion is 544.202 MeV, which is about 1,065 free electron masses. Therefore, 98 beta electrons and beta positrons, 49 pairs, must go into the bonds that hold the four pions together to form the K_π kaon.

The mass of K^0 is 497.611 MeV, which is about 974 free electron masses. Four π^0 particles have a mass of 1,056 free electron masses, making them 82 beta electrons and positrons or 41 pairs more than the K^0. The mass of a π^+, a π^- and two π^0 particles is 1,074 free electron masses, 100 free electron masses or 50 pairs, more than a K^0.

The four pions likely form a cluster of particles like those shown in Fig. 5.3. Each pion binds to the other three pions in the cluster, forming a tetrahedron. In this configuration, the four pions form six covalent-like bonds between them, by sharing beta electrons and positrons. These bonds produce the particle deficits that reduce the mass of the four free pions down to the mass of the kaon.

The covalent-like bonds form between any two pions within a K_π kaon. If two pions collide, beta particles from one pion could annihilate beta particles in the other pion. This would create a deficit of beta particles in the orbitals of both pions, causing the colliding orbitals in both pions to be unfilled.

To close the orbitals, each pion could share some of the betas in its unfilled orbital with the unfilled orbital of the other pion. If the collision annihilated a pair of betas from each pion, then each pion can share one of its remaining pairs with the other pion to fill its orbital. Because they then share betas, they are bound to each other.

Since the decay of a pion occurs at the central cluster, sharing particles within the shell should not affect the ability of the individual pion to decay. That is why some K_π kaons emit photons, resulting from π^0 decay, or lepton pairs from charged pion decays. Once a pion decay occurs within a kaon, the kaon most likely disintegrates.

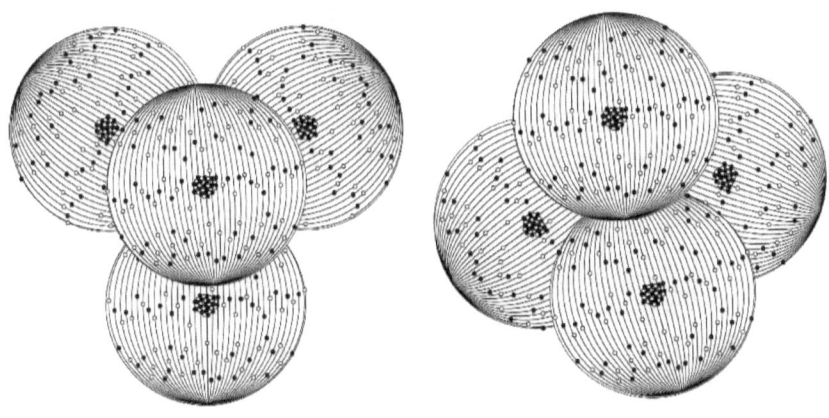

Fig. 5.3: Diagram of the K_π kaon model
Two views of the K_π kaon depicted as four pions bound together by six covalent-type bonds forming a tetrahedron. Depending on the kaon charge, the pions can be a variety of combinations.

Evidence of the multiple pions forming the charged K_π kaon is seen in Fig. 5.4. There, a K^+ kaon is shown decaying into two π^+ and one π^- particles. The kaon enters in a beam from the left and strikes an electron that veers it off its path. Shortly, it decays, sending two π^+ pions off in opposite directions and a π^- pion continuing in the direction the kaon was traveling. The identities of the three particles are confirmed by their decaying into muons.

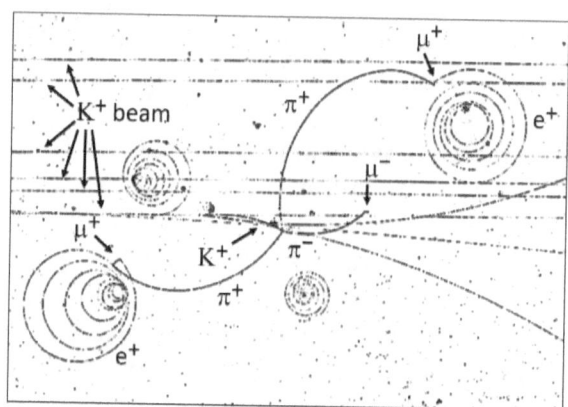

Fig. 5.4: Bubble chamber tracks of kaon decaying into three pions
A K^+ particle from a beam decays into two π^+ particles and a π^- particle.

One way to interpret the decay is that the kaon is a K_π kaon, made of a π^0, a π^- and two π^+ pions like the one shown on the left in Fig. 5.5. As the kaon travels along, the π^0 decays (middle). The energy from the decay separates the other three pions and launches them away from the point of the π^0 decay (right). No quarks, no gluons, no W boson as per Fig. 3.10, just four pions blown apart by the decay of one of them.

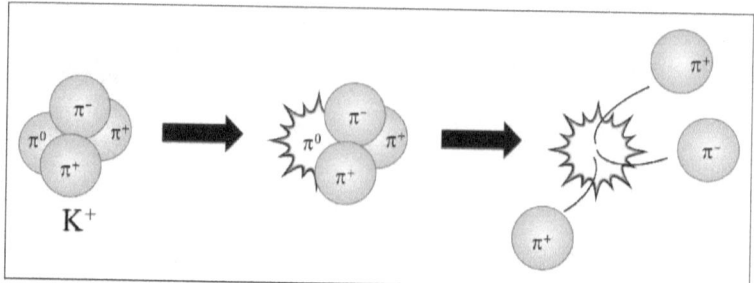

Fig. 5.5: Possible event sequence of kaon decaying into three pions
The pion cluster forming a K^+ kaon is blown apart into two π^+ particles and a π^- particle by the decay of its π^0 particle.

Whenever a kaon decays into one massive particle, either a muon or/and electron, and some neutrinos and/or photons, it was probably a K_μ kaon. The K_μ kaons cannot decay into multiple pions. If the kaon decays into more than one massive particle, be they pions, muons or electrons, then it was a K_π kaon.

The D mesons, both charmed (D) and strange (D_s), show signs of being made of possibly three, but likely four, kaons. The charged D meson mass of 1,869.3 MeV is about 3.8 times the mass of the kaon. Therefore, four kaons, with mass defect for binding, can easily make a D meson. Like the kaons, there appears to be a variety of D meson configurations made of various combinations of K_μ and K_π kaons.[75]

For example, about 4% of charged D mesons decay into a kaon, a pion, an electron and its antineutrino; while 3% of D mesons decay into a kaon, a pion, a muon and its antineutrino. These decay channels suggest these D mesons are made of at least one K_π meson and two K_μ kaons. The K_π kaon survives the decay. One of the K_μ kaons decays into a pion, and the other into a muon which sometimes decays into an electron.

In one to two percent of the D meson decays, three kaons emerge. These decays are likely from D mesons made of three K_π kaons that all survive the disintegration caused by a fourth K_μ decaying. Several prominent (>1%) decay channels show the D decaying into one or two kaons along with anywhere from two to four pions. In these cases, some of the K_π kaons probably break up into three pions.

The D_s mesons have a mass of 1968.5 MeV or about 3,852 free electron masses. This is almost exactly four times the mass of a charged kaon. Consequently, there is probably a pion in addition to the four kaons that make up the D_s mesons.[76]

5.3 A Path to the Proton

Based on electron-proton deep inelastic scattering data, the proton appears to be made of eight pions. So, the question arises: How did the eight pions come together to form a proton? One way it could have happened is for two K_π kaons to bind together.

The K_π kaons provide one of the best opportunities for nature to assemble the eight pions needed to form a proton. With each one being made of four pions, only two need collide and bond in order to form a particle containing eight pions. If one kaon is positive and the other neutral, together they likely form a proton.

Four pions can come together in seven ways to form kaons. Three: $(\pi^0 \pi^0 \pi^0 \pi^0)$, $(\pi^+ \pi^- \pi^0 \pi^0)$ and $(\pi^+ \pi^- \pi^+ \pi^-)$ form neutral kaons and four: $(\pi^+ \pi^- \pi^+ \pi^0)$, $(\pi^+ \pi^0 \pi^0 \pi^0)$ and their conjugates form charged kaons. The average bond strengths needed to hold these together to form kaons are 10.1, 8.6, 7.1, 10.0 and 8.5 MeV, respectively.

Two kaons can bind together in a variety of configurations. The diagrams in Fig. 5.6 show four of the more likely ones. The diagram labeled A has two pions joined together by one bond. This configuration is the easiest to form because only one pion from each kaon collides with the other to make it happen.

However, the mass of a charged and a neutral kaon is 991.3 MeV compared to the proton mass of 938.3 MeV. The bond holding the two kaons together would have to be 53 MeV. This is much larger than the bonds of 7 – 10 MeV that hold the pions together in the kaon. Consequently, it is probably not the proton configuration.

In diagram B, two pions from each kaon are bound together. This configuration probably happened a lot; however, probably not as often as that in diagram A. It is possible for one of the kaons in diagram A to swing around, causing one of its free pions to contact a free pion in the other kaon and form a configuration like that in diagram B. Naturally, two bonds are stronger than one; so, diagram B is more likely to survive in a hostile environment.

Unlike the configuration with one bond holding the kaons together, the two bonds holding them together would average 26.5 MeV each. At three times the average bond holding pions together in the kaons, this is probably not the configuration of the proton.

Three bonds formed in the collision of two kaons in diagram C. The three bonds make the particle very resilient to other collisions, relative to the prior two configurations. However, the three bonds still average 17.67 MeV, twice as much as the bonds between the pions in the kaons. This is probably not what the proton looks like.

Finally, the configuration in diagram D forms six bonds between the two kaons. This configuration forms a tightly packed cluster of the pions and probably is the most likely of the four to survive a harsh environment. The six bonds mean that the average bond strength is about 8.8 MeV, which is about the average strength of the bonds holding the pions together within the kaons. This makes it a good candidate for the proton configuration. However, at this point, there is no way to tell if the proton is any of these configurations.

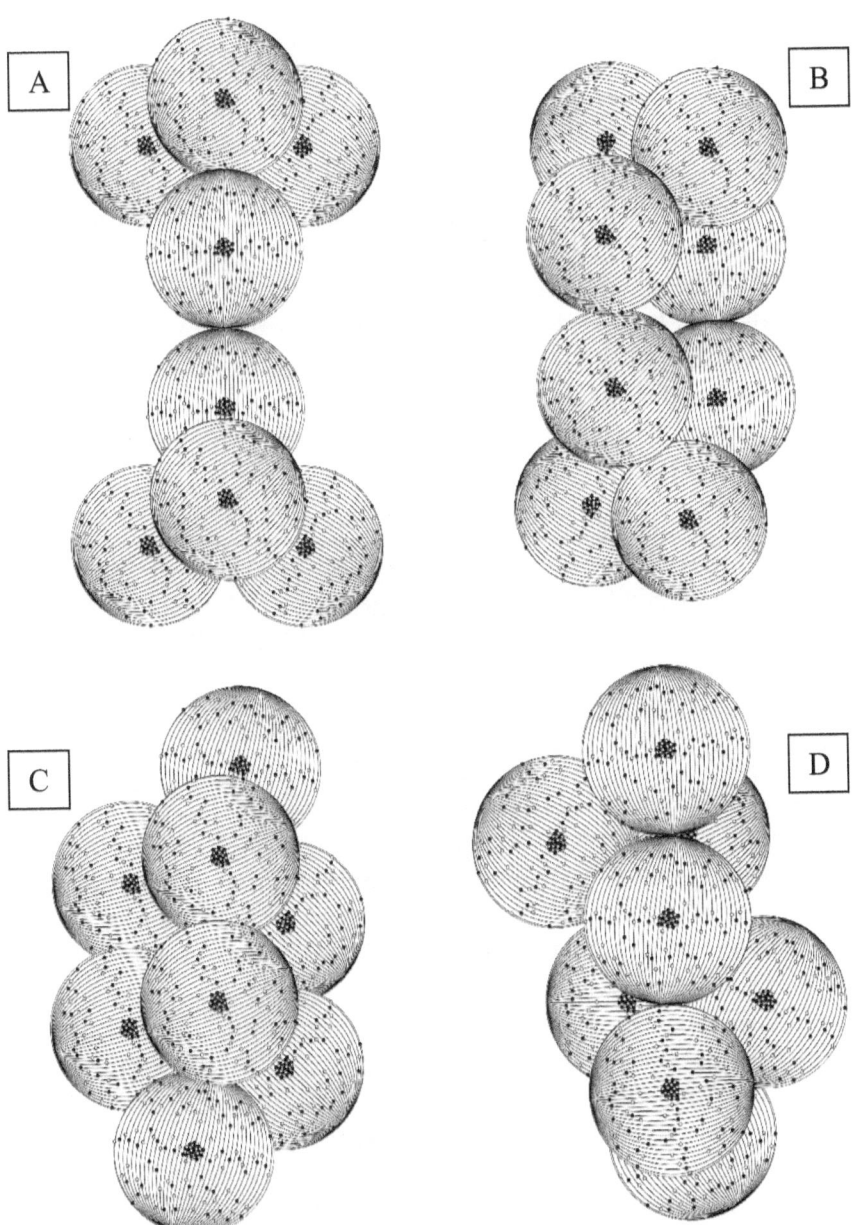

Fig. 5.6: Diagrams of two Kπ kaons bound together
It seems likely that the proton is made of a positive Kπ kaon and a neutral Kπ kaon bound together. Two bound Kπ kaons can form one (A), two (B), three (C), or six (D) bonds between them to possibly form a proton. The one with six seems the likely candidate for the proton.

The two positive and three neutral possible kaon configurations can come together to form six possible configurations of the proton. Therefore, there could be anywhere from one to six different types of proton. Each one would have the same mass and the same charge, but their pion content and distribution would be different.

Table 5.1 shows the combinations of pions that can possibly form a proton. Among the six, proton A2 and B1 contain the same set of pions distributed differently between the two kaons. Protons A3 and B2 also have the same pions arranged differently. The different types and arrangements of pions within the kaons forming the proton may have an effect on how the proton bonds with other nucleons.

Table 5.1: Possible Protons from Positive and Neutral Kaons

		Positive Kaons	
		A: $(\pi^+ \, \pi^0 \, \pi^0 \, \pi^0)$	**B:** $(\pi^+ \, \pi^- \, \pi^+ \, \pi^0)$
Neutral Kaons	**1:** $(\pi^0 \, \pi^0 \, \pi^0 \, \pi^0)$	**A1** $(\pi^+ \, \pi^0 \, \pi^0 \, \pi^0)$ $(\pi^0 \, \pi^0 \, \pi^0 \, \pi^0)$	**B1** $(\pi^+ \, \pi^- \, \pi^+ \, \pi^0)$ $(\pi^0 \, \pi^0 \, \pi^0 \, \pi^0)$
	2: $(\pi^+ \, \pi^- \, \pi^0 \, \pi^0)$	**A2** $(\pi^+ \, \pi^0 \, \pi^0 \, \pi^0)$ $(\pi^+ \, \pi^- \, \pi^0 \, \pi^0)$	**B2** $(\pi^+ \, \pi^- \, \pi^+ \, \pi^0)$ $(\pi^+ \, \pi^- \, \pi^0 \, \pi^0)$
	3: $(\pi^+ \, \pi^- \, \pi^+ \, \pi^-)$	**A3** $(\pi^+ \, \pi^0 \, \pi^0 \, \pi^0)$ $(\pi^+ \, \pi^- \, \pi^+ \, \pi^-)$	**B3** $(\pi^+ \, \pi^- \, \pi^+ \, \pi^0)$ $(\pi^+ \, \pi^- \, \pi^+ \, \pi^-)$

There may be no way to determine if all six combinations in Table 5.1 are actual protons. However, the bubble chamber photograph in Fig. 2.3 suggests that possibly four of them are. In that photo, a proton and an antiproton supposedly collide, shattering into four π^+ particles and four π^- particles. The diagram in Fig. 5.7 shows a drawing of the photo, labeling the particles resulting from the collision.

The fact that the particle-antiparticle collision did not result in an annihilation suggests the particles were not exact conjugates of each other. They had different numbers of charged and neutral pions.

It can be assumed that eight neutral pions also resulted from the collision, since two protons contain a total of 16 pions. The neutral pions did not show up because only charged particles leave tracks in the bubble chamber. Without a charge, they are invisible to the bubble chamber. Also, the π^0 lifetime is about 10^{-16} seconds. Consequently, even at the speed of light, they only travel 25 nm before they decay into one or two photons, which also do not leave tracks.

Ten combinations of the six particles in Table 5.1 and their anti-particles can produce eight charged and eight neutral pions from the proton-antiproton collisions. They are: $(A1,\overline{B3})$, $(\overline{A1},B3)$, $(A2,\overline{A3})$, $(\overline{A2},A3)$, $(A2,\overline{B2})$, $(\overline{A2},B2)$, $(A3,\overline{B1})$, $(\overline{A3},B1)$, $(B1,\overline{B2})$ and $(\overline{B1},B2)$.

It appears that, to get a proton-antiproton annihilation, each of the six particles formed in Table 5.1 has to interact with its conjugate particle. That is: $(A1,\overline{A1})$, $(A2,\overline{A2})$, $(A3,\overline{A3})$, $(B1,\overline{B1})$, $(B2,\overline{B2})$ and $(B3,\overline{B3})$. Because they contain the same set of pions just distributed among the kaons differently, $(A2,\overline{B1})$, $(\overline{A2},B1)$ and $(A3,\overline{B2})$, $(\overline{A3},B2)$ may also produce annihilations.

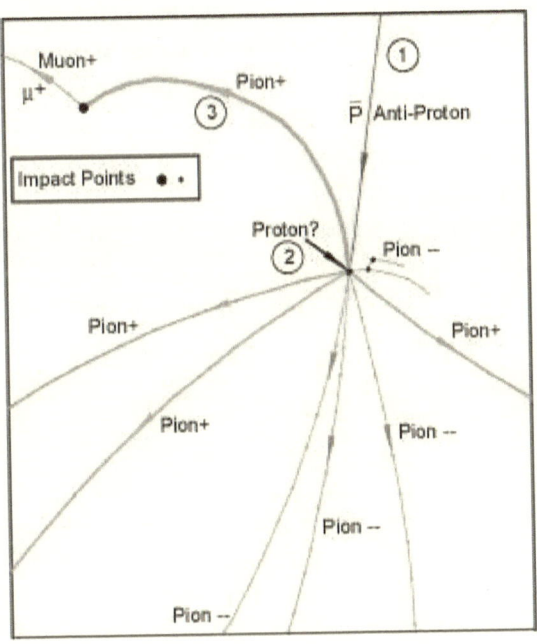

Fig. 5.7: Pions produced in proton-antiproton collision
Diagram of the photo from Fig. 2.3 identifying the particles created by the proton-antiproton collision. It shows four positive pions and four negative pions formed in the collision.

These analyses seem to confirm that the proton is made of a positive K_π and a neutral K_π. The two K_π mesons each contribute four pions to the structure of the proton, giving it the eight particles found by the electron-proton deep inelastic scattering. The fact that collisions of the antiparticles do not always result in annihilation tends to validate the idea that the six protons listed in Table 5.1 do exist. If so, this is another feature that the Standard Model failed to predict.

5.4 They Are All Related

According to the Standard Model, mesons are made of two quarks, whereas leptons are a completely different class of particle. Within the mesons, the pions are made of only up and down quarks, but kaons all contain strange quarks. This all gives the impression that these particles are significantly different from each other. However, the models offered here show that they do not have to be very different, at all. In fact, they are closely related.

The fact that the models of the kaon, pion and muon offered here are conveniently similar speaks to the efficiency of nature. Though the Standard Model has classified them as different entities; in fact, they are merely different phases of the same particle. Pions do not undergo a radical reconfiguration to become muons, their beta particle cores simply melt away to do so, as is the case for the kaons.

The Standard Model portrays mesons as particles made of quarks and leptons as particles not made of quarks to explain their different behaviors. The models here show that both families of particles are made of the same ingredients – beta electrons and positrons, but still exhibit the physics observed. In the end, the leptons are just descendants of the mesons.

6. Forcing In Bosons

In the 1950s, the idea caught root that radioactive decay, especially beta decay, may be caused by a force acting within a particle or nucleus. If so, the so-called "weak interaction" might be mediated by a set of force-carrying particles, like photons were thought to transmit the electromagnetic force. Consequently, two intermediate bosons, the W and the Z, were invented to model the decay of particles.

As was with the quarks, the relation of the W and Z to reality seemed secondary to their analytical appeal. Unnatural situations were ignored for the sake of a beautiful theory. The effort ultimately led to the combining of this theory of weak interactions with electromagnetic theory to form the theory of electroweak interactions.

However, that the W and the Z have mass breaks the symmetry needed to unify the two theories, which requires force-carrying particles be massless. To remedy this, Peter Higgs and others proposed that a field permeates space that gives particles that interact with it mass – the "Higgs" field. The carrier of this field is the Higgs boson.

Like other aspects of the Standard Model, the need for a theory of weak interactions disappears upon closer examination of what happens inside particles. With it goes the need for the W, the Z and the Higgs bosons.

6.1 Neutron Decay

Beta decay of a nucleus increases its atomic number by one without changing its mass number. Consequently, it was initially viewed as the decay of a neutron into a proton by emission of an electron, or

$$n \rightarrow p + e^-. \tag{6.1}$$

This suggests that the electrons emitted come from within the neutron with fixed energies. However, measurements of the electron energies from decaying sample showed they varied and were continuous. This led Pauli to hypothesize that another particle was emitted with electrons when beta decay occurred.[77] Fermi developed this idea, naming the particle the neutrino,[78] and equation 6.1 became

$$n \rightarrow p + e^- + \bar{\nu}. \tag{6.2}$$

This Fermi decay process is called a four-fermion interaction and is depicted by the diagram in Fig. 6.1.

In this theory of beta decay, the neutron does not disintegrate into a proton and an electron. Instead, an incoming neutrino, which is shown as an outgoing antineutrino, interacts with the neutron. This changes the neutron into a proton and causes the neutrino to become an electron.

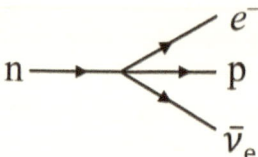

Fig. 6.1: Neutron decay via the Four-Fermion interaction
Diagram shows the incoming neutrino (outgoing antineutrino) interacting with neutron converting the neutron into a proton and the antineutrino into an electron.

This theory works well for low energies but breaks down at energies above 300 GeV because the relative neutrino cross section becomes greater than one. To remedy this shortcoming, in 1956 it was proposed that beta decay was caused by a weak interaction similar to the electromagnetic force. That is, a particle is responsible for producing the weak interaction like the photon causes electromagnetic interactions.[79]

However, unlike the photon, because it changes the charge of the decaying particle, the weak force particle must be charged. Also, because the decay occurs over a short range, within the nucleus of an atom, the particle had to be massive. From this, the W bosons were hypothesized.[80] With them, the neutron decay became the process shown in Fig. 6.2. The neutron emits a W^- boson to become a proton and the W^- decays into an electron and its antineutrino.

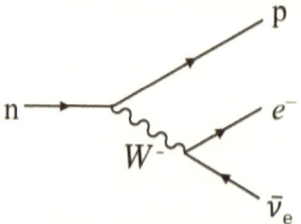

Fig. 6.2: Neutron decay via W^- boson interaction
Diagram shows the neutron decaying into a proton and a W^- boson. The W^- then decays into an electron and an electron antineutrino.

Even though, during the decays the electrons and the neutrinos appear to emanate from within the neutron; in both the above scenarios, the electron and the neutrino form outside the neutron. This was appealing because, if the electron decay energies were representative of their energies prior to the decay, according to the Heisenberg Uncertainty Principle, a container the size of a neutron, or even a nucleus, could not house a particle the size of an electron.

Another appealing aspect of this model was that muon decay, which also emits an electron and its antineutrino, could be modeled essentially the same way. The diagram in Fig. 6.3 shows the muon can emit a W^- boson to become its neutrino and the W^- could decay into the electron and its antineutrino.

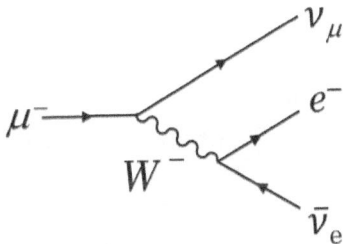

Fig. 6.3: **Muon decay via W^- boson interaction**
Diagram shows the muon decaying into a muon neutrino and a W^- boson. The W^- then decays into an electron and an electron antineutrino.

This interpretation of neutron decay essentially gave life to the W bosons. Of course, when adopted, no W bosons had been observed in nature. However, armed with a set of indirect indicators, a search was mounted, and the W proclaimed found in 1982 in high-energy $\bar{p}p$ collisions at CERN.[81]

Perhaps the W bosons do exist, but are they really the purveyors of particle decay?

6.2 W Boson – Jack of All Trades?

In the Standard Model, the W bosons are responsible for most of the particle decays observed. They can cause baryons to emit leptons and mesons, mesons to emit leptons and leptons to emit other leptons. They can be emitted by any type of fermion and can produce any fermion, as needed to affect a particle decay.[82] This versatility is believed possible because of its large mass, 80.38 GeV, which gives it the bulk it needs to produce almost any particle.

However, from earlier discussions of baryons, mesons and leptons, other mechanisms are available to explain the decay of these particles without the emission of a *W* boson. Mechanisms that align with the physical makeup of the particle.

For example, when a muon decays, it does not become a muon neutrino by emitting a *W* boson. In that scenario, a particle with a mass of about 105 MeV splits into two particles, one with a mass of 80 GeV and the other with virtually no mass. Even by Standard Model standards, this must seem far-fetched.

The muon model offered in chapter 4 gives a much more believable scenario for how the decay occurs. The muon is made of 103 beta electron-positron pairs and a valence beta particle orbiting a muon neutrino. These give it its 105 MeV mass.

Once the decay commences, the 103 pairs of beta electrons and beta positrons annihilate, freeing the muon neutrino. This leaves the unpaired valence beta particle in a pool of energy. The valence beta particle spawns a neutrino-antineutrino pair out of the energy, from which it captures the appropriate particle to become a free electron or positron. The leftover neutrino becomes a decay product.

In the latter, non-Standard-Model scenario, everything that happens during the decay does so with the mass and energy available before the decay. No supermassive particle need materialize, and the muon does not turn into a neutrino. No *W* boson is needed to make the decay happen.

The same is true of the pion decaying into a muon. There, 33 centrally located beta particle pairs of the 136 making up the pion annihilate. This provides energy for a valence beta electron or positron to spawn a muon neutrino-antineutrino pair. Once the neutrinos form, the remaining 103 beta particle pairs capture the neutrino, if the valence particle is an electron, or the antineutrino if it is a positron, to become a muon or antimuon. The remaining neutrino becomes the other half of the lepton pair. Again, no *W* bosons are needed.

According to the Standard Model, charged pions decay because the two quarks forming them annihilate each other creating a *W* boson (Fig. 6.4). That *W* boson then decays into a muon and its antineutrino. Again, a particle with a mass of 80 GeV is formed from the annihilation of two particles with a combined mass of 139 MeV. Of course, the energy mismatch is excused because the *W* only exists for 10^{-25} seconds. The Standard Model allows this. How convenient.

Fig. 6.4: Standard Model Pion Decay
Diagram shows the down quark and the anti-up quark annihilating, forming a W⁻ boson. The W⁻ then decays into a muon and a muon antineutrino.

In the Standard Model, quarks can emit W bosons to decay into other quarks, with the W boson decaying into lepton pairs or quark pairs (mesons). According to the Standard Model, when the neutron decays, one of its down quarks emits the W^- boson shown being emitted in Fig. 6.2. This transforms it into an up quark, changing the baryon from a neutron to a proton. In this instance, the W^- decays into a lepton pair, an electron and an electron antineutrino (Fig. 6.5).

The model of the neutron offered here is a proton containing two extra beta electrons and an extra beta positron. The positron annihilates one of the electrons creating energy that the other electron uses to spawn a neutrino-antineutrino pair. The electron then captures the neutrino to become a free electron and the antineutrino forms the other half of the lepton pair. Everything needed to cause the decay is contained within the neutron before the decay.

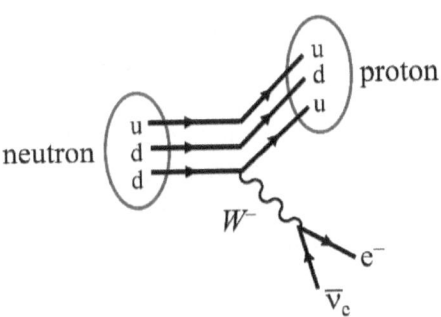

Fig. 6.5: Standard Model Neutron Decay
Diagram shows the neutron decaying into a proton and a W⁻ boson. The W⁻ then decays into an electron and an electron antineutrino.

When the K⁺ meson decays, the Standard Model has the *W*⁺ boson decaying into a pair of quarks. The anti-strange quark of the K⁺ meson emits a *W*⁺ boson to become an anti-up quark. The *W*⁺ boson decays into an up – anti-down quark pair, forming a pion.

The diagram on the left in Fig. 6.6 shows that that Standard Model K⁺ decay also involves the emission of a gluon, and ultimately results in the K⁺ decaying into three pions. Section 5.2 offered a more realistic scenario for this K⁺ decay, no *W* boson needed.

The diagram on the right in Fig. 6.6 shows a similar Standard Model decay of the D⁺ meson into a kaon and two pions, one of which forms from a *W*⁺ boson decay.

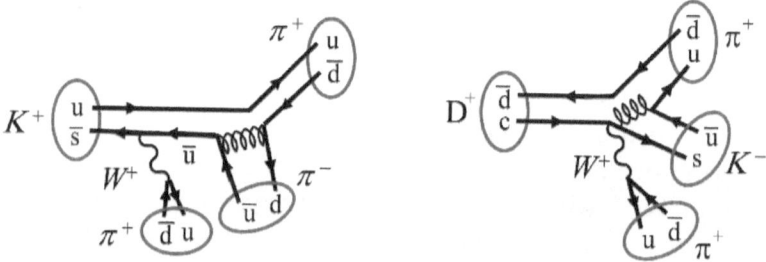

Fig. 6.6: Standard Model K⁺ and D⁺ Meson Decay
Diagrams show quarks emitting W⁺ bosons that decay into up – anti-down quark pairs that form pions.

Discussions in the last chapter indicated that pions are not particles formed outside a larger particle during decay. They are, instead, components of the greater particle before the decay. The discussions point out that, based on decay channels, at least two configurations of charged kaons exist.

One kaon configuration is a large cluster of beta electrons and beta positrons with 103 beta electron-positron pairs orbiting it. This one decays into a muon and its neutrino. The other configuration is a collection of four pions bound together. This one decays into one, two or three pions that existed within the kaon before it decayed.

Likewise, with D meson decay, it does not decay into pions and kaons because the D emits *W* bosons and gluons that ultimately form these particles. It decays into pions and kaons because it contains pions and kaons. The D meson is made of pions and kaons. Consequently, no *W* bosons (or gluons) are forming them during the decay. The pions and kaons are just escaping their confinement.

These examples strongly suggest that the *W* boson is not needed to explain particle decay. Particle decay is not caused by a "weak" force. It results from mass, energy and spatial transitions occurring inside the decaying particle. The mass and energy transitions are due to pair annihilation and pair production of particles forming the parent and daughter particles.

6.3 The Z Boson

The Standard Model casts the *Z* boson in a role similar to that of the *W* bosons, but the *Z* does not get as much airtime. The *Z* boson is the neutral mediator of the weak interaction in the Standard Model. Since they are neutral, they decay into particle-antiparticle pairs.[83]

One might expect them to show up whenever a particle decays into an electron and a positron (e.g., $\pi^0 \to e^+ e^-$) or a muon and an antimuon. However, a somewhat extensive literature search turned up few Feynman diagrams of particle-antiparticle pairs produced by *Z* bosons during particle decays. One example, $B^+ \to \pi^+ e^+ e^-$, is shown in Fig. 6.7.

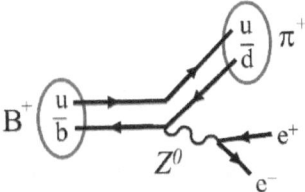

Fig. 6.7: One Standard Model B⁺ Decay Mode
Diagram shows a positive B meson decaying into a π^+ and a Z boson. The Z then decays into an electron and a positron.

In this decay, the anti-bottom quark, $\bar{b}$, with a charge of $+1/3$, emits a *Z* to become an anti-down quark, $\bar{d}$, which retains the $+1/3$ charge. In the Standard Model, the $\bar{b}$ supposedly emits a *Z* instead of a photon because of the large mass difference between the *b* quark and the *d* quark, about 4.5 GeV versus about 0.33 GeV.

Similar decays that likely also produce the *Z* boson are a B^+ meson decaying into a K^+ meson, an electron and a positron ($B^+ \to K^+ e^+ e^-$), also a B^+ meson decaying into a π^+ meson, a muon and an antimuon ($B^+ \to \pi^+ \mu^+ \mu^-$), and finally a B^+ meson decaying into a K^+ meson, a muon and an antimuon ($B^+ \to K^+ \mu^+ \mu^-$).[84] All these have low branching fraction, with the ones decaying into pions in the neighborhood of 10^{-8} and the kaon fractions on the order of 10^{-7}.

When the B^+ meson transforms into the K^+ meson, its anti-bottom quark, $\bar{b}$, becomes an anti-strange quark $\bar{s}$. Again, with an s quark mass of about 0.5 GeV, the large mass difference between it and the b quark likely requires the $\bar{b}$ quark to emit a Z to become an $\bar{s}$ quark in the Standard Model.

These decay modes along with ones that result in combinations of D, K and π mesons suggest that the B meson comes in a variety of configurations like the D and K mesons. One of those is the cluster of beta electron-positron pairs at its center along with a valence beta electron or positron, seen in the π and K_μ mesons (see chapter 5).

In large central clusters, as the beta electron-positron pairs annihilate, they can create enough energy so that a charged particle in the cluster can spawn a neutrino-antineutrino pair before the annihilation of the central core is complete. When this happens, one of the beta electrons not yet annihilated can capture the neutrino to become a free electron. Simultaneously, a beta positron can capture the antineutrino to become a free positron. This essentially creates an electron-positron pair without the Z boson. The emission of the free electron and positron confirm that the neutrinos were produced.

In the case of the B meson, if the electron and positron form before enough of the beta pairs in the cluster have decayed to form a kaon, a charged kaon may form. Or, the annihilations may continue until only enough clusters are left to form a charged pion. If they form after the kaon limit has passed, then only the pion can form.

All these decay modes are rare. Their scarcity alone calls into question the validity of the Z. To facilitate these decays, like the W, the Z appears out of nowhere, lasts 10^{-25} seconds, and has a mass of 91 GeV. These all make the Z seem more like an imaginary convenience than an actual particle, another modeling device (like quarks) created to explain observations in a way that fits the Standard Model.

According to the Standard Model, the Z boson decays into lepton-antilepton pairs about 10% of the time. The three leptons: electrons, muons and tau particles, occur about equally (3.33% each). About 70% of the time, Z bosons supposedly decay into quark-antiquark pairs. Particle jets are thought to indicate that the Z bosons have produced the quark-antiquark pairs. The last 20% of the Z boson decays are believed to be into neutrino-antineutrino pairs. These are assumed to occur based on energy and momentum conservation calculation, since neutrinos are not visible to the detectors.

All three of these types of decays can be explained without the emission of a Z boson. For particles decaying into lepton-antilepton pairs, charged pions and kaons offer a mechanism that does not require the formation and decay of a Z boson. The pions and the K_μ kaons have a cluster of beta electron-positron pairs at their center. When they decay, beta pairs in the cluster start annihilating. This creates a cloud of energy around and within the central cluster of pions or kaons.

Once enough annihilation energy is created, any beta electron or positron in the cluster can spawn a neutrino-antineutrino pair. Then a beta electron can capture the neutrino to become a free electron and a beta positron, the antineutrino, becoming a free positron. With that, the electron-positron pair is created.

It has been firmly established in earlier chapters that quarks do not exist. Consequently, Z bosons cannot decay into quark-antiquark pairs. The particle jets that supposedly signal the decay of a Z boson must have some other source. Caused by high-energy scattering, it seems reasonable to conclude the particle jets are the results of deep inelastic scattering. The particles in the jets are just fragments of the particles shattered in the collision.

It has been shown that both hadrons and mesons, particles made of quarks in the Standard Model, are actually collections of pions, electrons and positrons. An incident particle with enough energy could easily shatter the hadron or meson, unleashing a spray of particle fragments. A jet, if you will.

As for Z boson decays into neutrino-antineutrino pairs, a mechanism to produce these pairs from particles found within the decay particle has been discussed here and in previous chapters. Using the annihilation energy, valence beta electrons and positrons spawn electron neutrino-antineutrino pairs.

In instances involving the decay of charged particles, one of these particles is captured by a beta electron or positron to form a free electron or positron, and the other escapes the decaying particle intact. However, in neutral particles such as π^0 or K^0, the neutrino-antineutrino pair may form, but all the beta electron-positron pairs annihilate before the neutrino and antineutrino can be captured and become part of a free electron or free positron. Now, both the neutrino and the antineutrino can escape the decaying particle. This scenario can also occur in charged K_π mesons that contain π^0 components.

Like the *W* bosons, the *Z* boson appears to be an invention to fill the gaps of the Standard Model in explaining neutral decays. Instead of addressing how the particle-antiparticle pairs can form and emerge from within the decaying particle, the Standard Model defaults to the *Z* boson.

It pulls the *Z* boson literally out of the air and uses the *Z* to create what is needed to close the decay. Consequently, the *Z* boson is nothing more than a "black box" in the Standard Model. Particle-antiparticle pairs go in and particle-antiparticle pairs come out. Usually different ones, but sometimes the same.

6.4 No Need for the W and the Z

The models of the leptons, the baryons and the mesons call into question the purpose, and essentially the existence, of the *W* and *Z* bosons. The primary function of the *W* bosons is to bridge the gap between parent particles and their descendants during decays. In that role, *W* bosons are thought to materialize and subsequently decay into whatever was needed to facilitate the observed decay.

In many instances, *W* bosons were made to decay into quark pairs that formed mesons. However, having shown that mesons do not appear to be made of quarks, this function vanishes. The *W* bosons were also charged with decaying into the lepton pairs – a charged lepton and its conjugate neutrino. The models discussed in earlier chapters provide explanations for the appearance of leptons during baryon and meson decays without the need for *W* bosons.

Given the *W* boson supposedly exist for only 10^{-25} seconds, but for the claim that it has been found in experiments, one might wonder if the *W* exists, at all. It was proposed to give substance to the hypothesized weak interaction and make it somewhat analogous to the electromagnetic interaction. It supposedly carries the weak force the way photons carry the electromagnetic force.

However, the models presented here show that particle decay is not about the mediation of force. It is about the annihilation and production of pairs of particles. The pion does not emit a *W* boson that decays into a muon and a neutrino. Instead, 33 pairs of its beta electrons and positrons at the center of the pion annihilate. An unpaired beta forms a muon neutrino-antineutrino pair from the energy created by the annihilations. One of the neutrinos becomes the center of the decayed particle, making it a muon and the other escapes it.

The *W* bosons are also not needed to explain baryon decays into mesons. The decays are an indication that the mesons observed being emitted during the decays are components of the baryons before the decay. The baryon decays reveal that pions are a fundamental component of baryons. They are also components of larger mesons such as K, D and B mesons.

As for the *Z* boson, the neutral decays can also be explained without resorting to creating a force carrying particle. Like the *W* bosons, the *Z* boson was invented to treat particle decays that result in particle-antiparticle pairs as if they occurred due to a force. In the Standard Model, parent particles emit *Z* bosons that decay into q-$\bar{q}$, e^+- e^-, μ^+- μ^-, τ^+- τ^- or ν -$\bar{\nu}$ pairs. Since quarks do not appear to exist, the need for a *Z* to decay into a q- $\bar{q}$ pair goes away. This eliminates about 70% of the *Z* decays that allegedly occur in the Standard Model.

The ν -$\bar{\nu}$ pairs appear to form in a decaying particle when a beta particle interacts with energy from annihilating beta electron-beta positron pairs during the decay. One of the neutrinos is usually retained by the decaying particle to form a lepton and the other is emitted as a decay product. However, scenarios discussed in this chapter describe how the neutrino-antineutrino pairs are likely created and emitted by decaying particles.

The bottom line is, while it may be possible to create them, the *W* and *Z* bosons are not needed to explain radioactive decay of particles. The decay is not the result of the emission of a force-carrying particle. Consequently, no weak force or weak interaction exists (in nature).

Like the quarks, the weak interaction is just a device used to characterize observations of particle behavior within the framework of the Standard Model. In other words, the Standard Model set the parameters from which the theory of weak interactions has been developed, instead of radioactive decay setting the parameters from which the theory must be developed.

6.5 What About the Higgs?

To this point, it has been shown that gluons likely do not exist, and the *W* and *Z* bosons may also not exist; but if they do, are not needed to explain particle decay. Aside from the photons, which are seen all around all the time, only the Higgs boson remains of the Standard Model bosons. The Higgs boson is considered the final piece of the conformation that the Standard Model is valid.

Peter Higgs hypothesized the boson in 1964.[85] The search for it lasted over 40 years. In 2012, it was declared found among the debris of proton-proton collisions in the Large Hadron Collider (LHC).[86]

In the Standard Model, the Higgs boson is thought to be the carrier particle of the Higgs field. The Higgs field was hypothesized to provide a mechanism by which the W and Z bosons can have mass and still be force-carrying particles like the photon and the gluon.[87]

By having mass, the W and Z bosons, as force carriers, appear to break the symmetry that allows for the unification of the electromagnetic and weak interactions.[88] For the electroweak theory to work, they needed to be massless as are the photon and the gluon. The Higgs field supposedly provides a mechanism for particles that interact with it to have mass and those that do not, to be massless.

Per the Standard Model, all particles are inherently massless. However, some particles can interact with the Higgs field through absorption of the Higgs boson. Through this interaction, the particles acquire mass. Therefore, particles like the W and the Z are massless like the photon and the gluon and technically do not break the symmetry. However, they appear to have mass because of their interaction with the Higgs field. Photons and gluons supposedly do not interact with the Higgs field and do not acquire any mass.

All the massive particles in the Standard Model supposedly acquire their masses in this way. They interact to differing degrees with the Higgs field. Neutrinos, having very little mass, only slightly interact with it, but the tau particle interacts strongly with it to get its 1,776 MeV mass. The top quark, with a mass of 173 GeV, interacts the strongest of all Standard Model particles.

Of the 16 particles other than the Higgs boson in the Standard Model, only two, the W and the Z, need the Higgs mechanism to fit into the model. The Higgs mechanism, with its field and its boson, was invented so that the W and the Z, as massive force-carrying particles, would fit into the Standard Model. The W and Z, themselves, were invented so that radioactive decay could be characterized as the emission of a particle by a weak force, like photons in the electromagnetic force. The weak interaction was invented to address the breakdown of the four-fermion neutron decay at energies greater than 300 GeV. Finally, the four-fermion interaction was invented to address the continuous energy spectrum electrons emit during beta decay.

So, in a sense, the Higgs boson is the Standard Model's way of addressing the continuous decay spectrum of the electron emitted during beta decay. But more specifically, it was devised so that the *W* and *Z* bosons could have mass and still fit into the electroweak interaction. That way, beta decay and neutral decays could be characterized as *W* and *Z* emissions.

It has been shown earlier in this chapter that the beta and neutral decays can be explained without *W* and *Z* bosons. They are not needed to model these decays, and likely may not even exist. Consequently, there is no need for the Higgs field for the *W* and *Z* to interact with to acquire mass and no need for a Higgs boson to facilitate the interaction.

As for producing the mass of the fermions in the Standard Model, it appears that the six quarks do not exist, and the three charged leptons are not fundamental. Only the beta electron (and positron) and the neutrinos still appear to be fundamental and have any mass. The only alleged force-carrying particle left, the photon, is massless. Therefore, there is no reason why these massive particles cannot possess their mass outright, without the aid of a Higgs field.

7. Neutrino Mystery

Neutrinos have proven to be very elusive particles for reasons other than their weak interaction with matter. For years, discrepancies between the number of expected neutrinos arriving at Earth from the Sun and the measured number produced the so-called *solar neutrino problem*. However, the problem was eventually declared resolved by the discovery of neutrino oscillation.

7.1 Missing in Action

According to the Standard Solar Model,[89] the energy of the Sun is produced by thermonuclear fusion of light nuclei in its core. The predominant reaction thought to take place is the fusion of two protons into a deuteron (^{2}H). Several of these reactions along with some decaying isotopes (shown in Table 7.1) are thought to produce neutrinos during their course.[90]

Table 7.1 Neutrino producing reactions in the core of the Sun

Type	No.	Label	Reaction	ϕ_v (cm^{-2}s^{-1})	E_v (MeV)
Fusions	1	pp	$p + p \rightarrow {}^2H + e^+ + v_e$	5.95 x 10^{10}	≤ 0.42
	2	pep	$p + e^- + p \rightarrow {}^2H + v_e$	1.40 x 10^8	1.44
	3	hep	$^3He + p \rightarrow {}^4He + e^+ + v_e$	9.30 x 10^3	≤ 18.77
	4	^7Be	$^7Be + e^- \rightarrow {}^7Li + v_e$	4.77 x 10^9	0.38, 0.86
Decays	5	^8B	$^8B \rightarrow {}^8Be^* + e^+ + v_e$	5.05 x 10^6	< 15
	6	^{13}N	$^{13}N \rightarrow {}^{13}C + e^+ + v_e$	5.48 x 10^8	≤ 2.22
	7	^{15}O	$^{15}O \rightarrow {}^{15}N + e^+ + v_e$	4.80 x 10^8	≤ 2.75
	8	^{17}F	$^{17}F \rightarrow {}^{17}O + e^+ + v_e$	5.63x 10^6	≤ 2.76

In 1964, John Bahcall did a calculation to predict how many of the neutrinos created in the Sun should reach Earth.[91] Because they are absorbed and reemitted over and over again, photons produced in its core could take millions of years to escape the Sun. However, since they barely interact with matter, the neutrinos formed in Sun's core pass right through it. Traveling at near light speed, they reach Earth in about eight minutes after they form. The neutrino fluxes and energies he calculated were similar to those shown in Table 7.1.

In 1965, Raymond Davis began constructing a facility to detect solar neutrinos in the Homestake Gold Mine in South Dakota.[92] The detector was a large tank filled with the liquid perchloroethylene (C_2Cl_4). It used the reaction $^{37}Cl + \nu_e \rightarrow {}^{37}Ar + e^-$ to detect the neutrinos. The ^{37}Ar is a radioactive gas with a half-life of 35 days. By periodically collecting the ^{37}Ar gas from the tank and counting the disintegrations for 200 – 250 days, an estimate of the solar neutrino flux could be determined.

The ^{37}Cl is sensitive to neutrinos with energies > 0.81 MeV; which, from Table 7.1, means that it could not detect the neutrinos from the *pp* reaction. Using assumptions from the Standard Solar Model, Bahcall determined the number of neutrinos the ^{37}Cl detector should see to be about 7.6 solar neutrino units (SNU), where a SNU = 10^{-36} events per target atom per second. [93]

Davis started collecting data in 1967. After 14 years of collecting data, the solar neutrino flux measurement from the ^{37}Cl detector was 1.8 SNU, about ¼ the prediction.[94] With 20 more years of data and some refinements in the data analyses, the measurement increased to 2.56 SNU, a factor of three lower than the model calculation.[95]

According to the calculation, the bulk of the expected neutrinos, 5.9 SNU (~76%), come from the ^8B reaction (see Table 7.2), with the ^7Be reaction contributing another 1.15 SNU (~15%). After reviewing the calculation and the measurement, it was concluded that the measurement was somehow deficient, not the calculation used.

Beginning in May 1991, experiments to measure the solar neutrino flux using gallium were performed at Laboratori Nazionali del Gran Sasso (LNGS) in Italy. The GALLEX (Gallium Experiment)[96] detector used 100 tons of gallium chloride ($GaCl_2$) to take advantage of the reaction $^{71}Ga + \nu_e \rightarrow {}^{71}Ge + e^-$.

The ^{71}Ge is radioactive, with a half-life of 11 days. After a period of exposure, the ^{71}Ge was extracted from the $GaCl_2$ and counted to determine the number of reactions that had occurred. This was similar, in principle, to the $^{37}Cl(\nu_e, e^-)^{37}Ar$ method of Davis at the Homestake detector. However, the $^{71}Ga(\nu_e, e^-)^{71}Ge$ reaction is sensitive to neutrinos with energies as low as 0.233 MeV, which means it can detect neutrinos from the *pp* reaction that the $^{37}Cl(\nu_e, e^-)^{37}Ar$ reaction cannot. In 1997, the GNO (Gallium Neutrino Observatory) replaced GALLEX and continued its mission.[97]

A calculation using the Standard Solar Model was done to determine what neutrino flux gallium detectors should see and predicted 129 SNU.[98] After 123 runs (GALLEX – 65 and GNO – 58), GALLEX/GNO measured an average of 69.3 SNU, about 54% of the flux predicted by the model.

Another gallium experiment, SAGE (Soviet-American Gallium Experiment), independent of GALLEX/GNO, was conducted in the Baksan Laboratory in Russia from 1990 to 2001.[99] SAGE used a gallium-germanium neutrino telescope made of 50 tons of gallium liquid metal in seven chemical reactors. After a four-week exposure, the ^{71}Ge is chemically extracted and counted to determine the neutrino flux experienced. For 92 runs over 12 years, SAGE measured an average of 70.8 SNU, about 55% of the predicted flux and very close to the GALLEX/GNO value of 69.3 SNU.

The two independent measurements validated their results placing the neutrino flux ^{71}Ga sees at 70 SNU. However, the calculation thinks that ^{71}Ga should see 129 SNU. With independent measurements corroborating each other but in severe disagreement with the model calculation; again, the calculation appears to be flawed, not the measurements. However, the measurements were suspected.

7.2 Neutrino Oscillations?

In 1957, Bruno Pontecorvo proposed that neutrinos and antineutrinos might oscillate between each other.[100] By 1967, two flavors of neutrinos, ν_e and ν_μ, were known to exist and Pontecorvo extended his proposal to electron and muon neutrinos ($\nu_e \leftrightarrow \nu_\mu$).[101] Initially, the idea was not well received. However, given the results of the Homestake neutrino detector, by 1970, in a manner similar to Gell-Mann's quarks, Pontecorvo's neutrino oscillations were embraced, and an intense effort to validate them undertaken. Over the next 30 years, several experiments would be performed, all designed to show that neutrinos oscillate from electron to muon to tau flavors and back.

Super-Kamiokande,[102] a detector 1 km underground in Japan, used 22,500 tons of ultra-pure water surrounded by 11,146 photomultiplier tubes to detect neutrinos with energies > 7 MeV, essentially ^{8}B neutrinos. Neutrinos scatter off electrons in the water, causing them to give off Cerenkov radiation, which was immediately detected by the photomultiplier tubes. The recoil of the electrons during the scatters also showed the direction from which the neutrinos came.

In 1995, it was claimed that Super-Kamiokande showed that atmospheric neutrinos, the neutrinos formed by cosmic rays in Earth's upper atmosphere, oscillate.[103] Cosmic rays produce charged pions that decay into negative and positive muons and muon neutrinos and antineutrinos. The muons subsequently decay into muon neutrinos and antineutrinos, electrons and positrons and electron neutrinos and antineutrinos.

Super-Kamiokande surveyed the electron and muon neutrinos coming from directly above it through the atmosphere, directly below it through the Earth, and from several angles between the two. The data suggested that, while there appeared to be no indication of electron neutrinos oscillating, the muon neutrinos appeared to be oscillating with tau neutrinos, or possibly with so-called sterile neutrinos, as they passed through the Earth.

In 1998, the Sudbury Neutrino Observatory (SNO) was built in Canada to determine if solar neutrinos oscillated. It is a transparent acrylic sphere 12 m in diameter, containing 1,000 tons of ultra-pure D_2O, buried 2 km underground. The sphere is surrounded by 9,456 inward-looking photomultiplier tubes to detect photons created inside it and 91 outward-looking photomultiplier tubes to monitor muons from cosmic rays that enter the detector. The detector is inside a geodesic sphere about 18 m in diameter, containing 7,000 tons of ultra-pure H_2O to support it and shield it from natural radiation.

The SNO detector used D_2O to detect neutrinos using three different reactions. [104] In the first reaction called charged current (CC) reaction, a neutron in the deuterium nucleus absorbs an electron neutrino. The excited neutron subsequently emits an electron to become a proton ($v_e + d \rightarrow p + p + e^-$). The emitted electron gives off Cerenkov radiation that is detected by photomultiplier tubes. This reaction only sees electron neutrinos.

The second reaction, elastic scattering (ES), can detect any flavor neutrino, but favors electron neutrinos. The neutrino scatters off an electron in the D_2O ($v_x + e^- \rightarrow v_x + e^-$), like in the H_2O in Super-Kamiokande. This causes the electron to give off Cerenkov light.

The third reaction is called a neutral current (NC) reaction. Any flavor neutrino can interact with a deuteron (^2H nucleus), causing the nucleus to break up into a proton, a neutron and a neutrino ($v_x + d \rightarrow p + n + v_x$). The appearance of the proton signals the detection of a neutrino, although the type is not known.

In 2002, Sudbury claimed to show that the neutrinos coming from the Sun also oscillate.[105] It claimed that the CC reaction, which can detect only electron neutrinos, saw a flux of 1.76×10^6 cm^{-2} s^{-1}. Additional neutrino flux detected by the ES and NC reactions, 3.41×10^6 cm^{-2} s^{-1}, was assumed to consist of muon and tau neutrinos. It detected a total solar ^{8}B neutrino flux of 5.09×10^6 cm^{-2} s^{-1} compared to 5.82×10^6 cm^{-2} s^{-1} predicted by the calculation.[106] The 1.76×10^6 cm^{-2} s^{-1} electron neutrino flux from the CC reaction is about ⅓ of the total neutrinos detected.

As a result of the SNO determinations, solar neutrino oscillations were declared the reason why the Homestake neutrino detector only saw ⅓ of the neutrinos expected. Like the quarks before them, once allegedly revealed, physicists quickly declared neutrino oscillations a real phenomenon, and the solar neutrino problem resolved. However, measuring only ⅓ the expected electron neutrino flux at Sudbury may align with the Homestake measurement, but it does not explain the results of the two gallium experiments.

7.3 What's the Real Problem?

Table 7.2 shows the Standard Solar Model calculated distributions of neutrino fluxes, in SNU, the chlorine (ϕ_{Cl}) and gallium (ϕ_{Ga}) detectors see, for each of the neutrino producing reactions thought to occur in the core of the Sun. The actual fluxes (ϕ_v) predicted by the model and the neutrino energies for each reaction are also shown.

Table 7.2 Neutrino fluxes predicted by the Standard Solar Model

Label	E_v (MeV)	ϕ_v (cm^{-2}s^{-1})	ϕ_{Cl} (SNU)	ϕ_{Ga} (SNU)
pp	≤ 0.42	5.95×10^{10}	0.0	69.6
pep	1.44	1.40×10^8	0.2	2.8
hep	≤ 18.77	9.30×10^3	0.0	0.0
^7Be	0.86, 0.38	4.77×10^9	1.15	34.4
^8B	< 15	5.05×10^6	5.9	12.4
^{13}N	≤ 2.22	5.48×10^8	0.1	3.7
^{15}O	≤ 2.75	4.80×10^8	0.4	6.0
^{17}F	≤ 2.76	5.63×10^6	0.0	0.1
Total			7.75	129.0
Measured			2.56	70.0

While the ^{37}Cl detector sees about 33% of the electron neutrinos predicted by the Standard Solar Model, consistent with the Sudbury result, the ^{71}Ga detectors see about 54% of the predicted electron neutrinos. If neutrino oscillation indicates that only 33% of the electron neutrinos from the Sun make it to Earth as electron neutrinos, then how does the ^{71}Ga detectors see 54%? With the ^{37}Cl and ^{71}Ga detectors seeing such different electron neutrino fluxes, clearly something is amiss.

There seems to be at least three possible sources of the discrepancy between the Homestake ^{37}Cl and the GALLEX/GNO/SAGE ^{71}Ga measurements. Either the measurements are wrong, the Standard Solar Model is wrong, or the calculations of the expected detector fluxes are wrong. In each case, there could be some unknown phenomenon not accounted for.

Regarding the measurements, the ^{37}Cl experiment ran for nearly 30 years and produced fairly consistent results over that period. Continual scrutiny of the process makes an undetected flaw in it unlikely. In the case of the ^{71}Ga experiments, two completely independent experiments using different techniques produced essentially the same result. The likelihood that both experiments would make the same mistakes in assessing their results seems remote.

As for the model, while there is no way to directly probe the Sun to determine its structure, according to helioseismology, the current Standard Solar Model is reasonably accurate.[107] Helioseismology analyzes waves propagating on the surface of the Sun to determine its internal structure and behavior. Based on the sound speed determined at the center of the Sun, its core is fusing hydrogen into helium as described in the Standard Solar Model. Consequently, helioseismology validates the neutrino production modeled via the various core reactions in Table 7.1. This appears to eliminate the model as the source of the discrepancies.

With the measurements and the model ruled out as sources of the discrepancies, the calculations become the prime suspects. The two main components of the calculations are the fluxes and the cross sections. The same set of fluxes is used by all the calculations, so any problem with them should vanish in comparison. And although, a unique set of cross sections is developed for each detector medium, according to the literature, those cross sections have been scrutinized at least as much as the experimental results, if not more.

A possibility one might consider is whether something about the detector affects the outcome of the measurement. For example, perchloroethylene is a liquid, with four chlorine atoms, each attached to a carbon atom. On average, one of the four chlorine atoms is a ^{37}Cl atom. When the ^{37}Cl nucleus captures a neutrino, it is transformed into a ^{37}Ar nucleus, and the assumption is that it disengages from the former C_2Cl_4 molecule, becoming a gas atom.

However, it may be that some of the ^{37}Ar atoms converted by the neutrinos remain attached to the former C_2Cl_4 molecule forming C_2Cl_3Ar. This molecule would likely remain part of the liquid mass, never making it to the radiation detector for measurement. In such a case, a factor must be applied to the flux calculation to account for the fraction of ^{37}Ar atoms remaining in the liquid, not counted.

According to his biography, Davis considered this possibility and devised a test to check it using C_2Cl_4 laced with radioactive ^{36}Cl, although no reference for the test was given.[108] The beta decay of ^{36}Cl into ^{36}Ar supposedly mimicked the neutrino-induced transmutation of ^{37}Cl into ^{37}Ar. He found that all the ^{36}Ar could be accounted for in the counting, indicating that all the ^{36}Ar atoms broke free of the former C_2Cl_4 molecules once they formed.

Another possibility is that the data from the detector is being misinterpreted. In both the ^{37}Cl and ^{71}Ga experiments, the final flux determination depends on counting the radioactive disintegrations of the product formed from interactions with neutrinos. Those counts are then deciphered into the total number of neutrino interactions that occurred during the run.

The literature indicates that much thought and care went into counting both the ^{37}Ar and ^{71}Ge samples. The likelihood of some error occurring seems small. That leaves interpreting the results of the counting. Again, the process for converting the counts to fluxes appears to be technically sound. However, there may be one aspect of this exercise that is so subtle that it has escaped all scrutiny, so far.

7.4 What Did Sudbury Really See?

From the Sudbury experiment, when an electron neutrino interacts with a deuteron, two things can happen. It can be absorbed and reemitted, causing the deuteron to break up into a proton and a neutron – the NC reaction; or it can be absorbed and reemitted as an electron, causing the deuteron to split into two protons – the CC reaction.

In both cases, the deuteron absorbs the electron neutrino. The question is, what causes the reaction to go charged or neutral? The answer may be found by using the models of the nucleons and electron described in chapters 2 and 4.

From Table 7.1, the *pp* reaction that forms a deuteron in the core of the Sun is $p + p \rightarrow {}^2H + e^+ + \nu_e$. What appears to happen is that sometimes when the two protons collide forming a diproton, an electron neutrino – antineutrino pair is formed. Once formed, a beta positron in the diproton captures the antineutrino to become the free positron. Then the positron and the leftover electron neutrino exit the nucleus, leaving a deuteron. This reaction proceeds very slowly within the Sun's core because for it to happen, the neutrino – antineutrino pair must form during the collision of the two protons.

When the beta positron is removed from the proton, it leaves the proton with the same number of beta positrons as beta electrons, rendering it neutral. This makes what was a diproton, a deuteron. It has one charged nucleon and one neutral one.

An electron neutrino with enough energy that interacts with a deuteron will cause it to separate into two nucleons. The neutrino can interact with either a beta electron or beta positron in either nucleon of the deuteron. This sets up four possible scenarios for the interaction: the neutrino can interact with a beta positron in the charged nucleon; it can interact with a beta electron in the charged nucleon; it can interact with a beta positron in the neutral nucleon; or the neutrino can interact with a beta electron in the neutral nucleon.

Since electron neutrinos only couple with beta electrons to form free electrons, interactions with the beta positrons only cause the neutrinos to scatter after separating the positive and neutral nucleons. If the electron neutrinos interact with beta electrons in the deuteron, they can form the free electrons seen emanating from the CC reactions. However, the free electron is only emitted when the deuteron separates into two protons. This indicates that the neutrino can only couple with a beta electron in the neutral nucleon to form the free electron. Apparently, the neutral nucleon will give up a beta electron to become a +1 nucleon, but the +1 nucleon will not give up one to become a +2 nucleon, which does not occur in nature.

Of the four possible scenarios the neutrino interaction with the deuteron can take, only one can produce a free electron. Assuming all four scenarios have an equal probability of occurring, one would expect the one forming the free electron to occur ⅓ as many times as the ones just scattering the neutrino. This explains why the Sudbury NC reaction saw three times as many neutrinos as its CC reaction.

The NC reaction was not seeing all the neutrinos, it was only seeing the electron neutrinos that interacted with either the beta positrons in the deuterons or beta electrons in the charged nucleons of the deuterons. Consequently, both the CC and the NC reactions are seeing only electron neutrinos and the total number of electron neutrinos Sudbury saw is the sum of the CC and NC fluxes, 6.85×10^6 cm^{-2} s^{-1}. The fact that the NC reaction saw three times the neutrinos the CC reaction did indicates it saw electron neutrinos.

7.5 What Did the Gallium Detectors See?

Given the outcome of the reanalysis of the Sudbury data, it is useful to revisit the results of the other neutrino detectors. The two gallium detectors, GALLEX/GNO and SAGE, both reported about 55% of the expected neutrino flux from the Sun. Those detectors relied on the neutrinos interacting with ^{71}Ga to form radioactive ^{71}Ge via the reaction ^{71}Ga + $\nu_e \to {}^{71}$Ge + e^-.

The ^{71}Ga nucleus has 71 nucleons, of which, 31 are positive and 40 are neutral. The ^{71}Ge nucleus also has 71 nucleons, but 32 of them are positive. From the reaction, like the ν_e + d reaction in SNO, it appears the neutrino interacts with a beta electron in the ^{71}Ga nucleus to form a free electron. Once the free electron exits the nucleus, it leaves it with 32 positive nucleons, making it ^{71}Ge. However, unlike the deuteron, where its two nucleons are held together by only one bond, each nucleon in the ^{71}Ga nucleus is held in the nucleus by multiple bonds. Because of this, the neutrino cannot separate a nucleon from the nucleus by breaking only one bond.

As is the case with the deuteron, the electron neutrino can interact with a beta electron or beta positron in the ^{71}Ga nucleus. When it interacts with a beta positron, it just scatters off it because it cannot couple with it to form a free positron. Only the electron antineutrino can do that. Since the neutrino cannot breakup the ^{71}Ga nucleus, there is no indication that it scattered off it. Unlike this case for the deuteron, evidence of the neutrino interaction cannot be seen. When the neutrino interacts with a beta electron in ^{71}Ga, it can create a free electron and form radioactive ^{71}Ge. This interaction can be seen.

Assuming the neutrino can interact with any nucleon in the ^{71}Ga with equal probability, it can interact with 31 positive nucleons and 40 neutral nucleons. If, like in the deuteron, only the neutral nucleon will give up a beta electron to the electron neutrino to form a free electron, then the neutrinos will also scatter off the beta electrons in the positive nucleons, leaving no sign of the interaction. Consequently, only 40 of the 71 possible interactions can be registered by the detector, 56%. This is essentially the same fraction of the predicted electron neutrino flux the two gallium detectors saw. So, it seems the gallium detectors were seeing the full solar flux as electron neutrinos but could only show 56% of it.

7.6 The Same, but Different

Like the gallium detectors, the chlorine detector relies on the electron neutrino to convert a stable isotope, ^{37}Cl, into radioactive one, ^{37}Ar, by coupling with a beta electron to form a free electron. Since the reaction, ^{37}Cl + v_e → ^{37}Ar + e^-, is similar to the ^{71}Ga reaction, one might expect a similar measurement result. There are 37 nucleons in ^{37}Cl and 20 of them are neutral. Therefore, the detector should see 20 out of every 37 interactions, or 54% of the predicted solar neutrino flux. However, the Homestake detector only measured about 33% of the predicted flux. What is causing this discrepancy?

The flux of electron neutrinos the detector sees during an exposure period is related to how much ^{37}Ar is produced in the detector. This is determined by collecting the ^{37}Ar gas from the detector and counting the radioactive disintegrations it produces in a counter. The general expression used to determine the ^{37}Ar production rate, p, from the number of ^{37}Ar decays the counter counts, N_c, is

$$p = \frac{N_c \lambda}{1 - e^{-\lambda t_{exp}}} = N_c \lambda',$$

where λ is the ^{37}Ar decay constant and t_{exp} is the exposure time of the ^{37}Cl to the neutrinos. This indicates that the number of ^{37}Ar nuclei produced is essentially proportional to the ^{37}Ar decay constant λ', which is inversely proportional to its half-life. Therefore, if the half-life for the neutrino-induced ^{37}Ar nuclei is shorter than that of the proton/deuteron induced ^{37}Ar nuclei; then its decay constant would be larger, and the calculated ^{37}Ar production rate would be greater.

When ^{37}Ar is created in the laboratory by bombarding ^{37}Cl with either a proton or a deuteron, the mass of the resulting ^{37}Ar nucleus, at 36.956690 u, is greater than the ^{37}Cl nucleus mass of 36.956577 u. The collision increases the mass of the nucleus. This ^{37}Ar nucleus has a half-life of about 35 days.

When ^{37}Ar is created by ^{37}Cl interacting with an electron neutrino, the neutrino removes one of the beta electrons from the nucleus, which should make the mass of the resulting ^{37}Ar nucleus less than that of the ^{37}Cl nucleus. The ^{37}Ar nucleus from neutrino interaction with ^{37}Cl is different from the ^{37}Ar created from ^{37}Cl in the lab. This begs the question, do the different configurations of ^{37}Ar decay at different rates? If so, then the neutrino flux value determined using the lab ^{37}Ar half-life is probably incorrect.

When a neutrino interacts with a ^{37}Cl nucleus, it is absorbed, and an electron is emitted by the nucleus. Losing a charge of -1 leaves the nucleus with a charge of +18, making it ^{37}Ar. Studies of ^{37}Ar produced in laboratories found it has a half-life of 35 days and decays by electron capture back into ^{37}Cl, but is it the same as this ^{37}Ar?

Consider ^{37}Ar produced from ^{37}Cl + p → ^{37}Ar + n. What appears to happen here is that a proton collides with a ^{37}Cl nucleus, knocking out and taking the place of one of its neutrons. The nucleus retains its mass number of 37, but its atomic number increases from 17 to 18, making it a ^{37}Ar nucleus. However, closer analysis reveals that this is likely not the case.

The mass of a ^{37}Ar nucleus at 36.956690 u is greater than the ^{37}Cl nucleus mass of 36.956577 u. Since the proton mass of 1.007277 u is less than the neutron mass of 1.008665 u, if it replaced a neutron in ^{37}Cl to form ^{37}Ar, the resulting ^{37}Ar mass would likely be less than the original ^{37}Cl mass and less than the measure ^{37}Ar mass.

What probably happens is that during the collision between the proton and the ^{37}Cl nucleus, the proton transfers one of its beta positrons to the ^{37}Cl nucleus. Assuming the mass of the beta positron is equal to that of a free electron, 0.000549 u, the transfer would increase the mass of the nucleus to 36.957126 u less some energy to bind the beta positron, and its charge to +18, making it ^{37}Ar.

Having lost one of its beta positrons, the once proton is now a neutral particle with a mass near, but less than a neutron. However, because it has no charge, its mass cannot be measured directly. Therefore, for all practical purposes, it serves as a neutron.

The other ^{37}Ar producing reaction that involves the ^{37}Cl nucleus, ^{37}Cl + d → ^{37}Ar + 2n, also appears to transfer a beta positron to the ^{37}Cl nucleus. This renders the deuteron neutral, causing it to split into two neutral neutron-like particles. If the deuteron, with a mass of 2.013553 u, replaced two neutrons (2.017330 u) in the ^{37}Cl nucleus, the resulting ^{37}Ar nucleus mass would be about 36.952800 u, much less than the measured ^{37}Ar mass. This likely cannot happen.

Therefore, the ^{37}Ar production modes that involve ^{37}Cl nuclei appear to give the ^{37}Cl nucleus an extra beta positron during the reaction. This increases the mass of the ^{37}Cl nucleus by 0.000325 u, one beta positron mass less some binding energy, and increases the charge of the nucleus by 1 from +17 to +18, making it ^{37}Ar.

In neutrino-induced ^{37}Ar production from ^{37}Cl, the ^{37}Cl nucleus appears to capture the neutrino, then emit an electron to become ^{37}Ar (v_e + ^{37}Cl → ^{37}Ar + e^-). What apparently happens is that the neutrino, while passing through the ^{37}Cl nucleus, couples with one of its beta electrons to form a free electron. Once formed, the free electron either escapes, or is ejected, from the nucleus.

Like in the production via the proton and deuteron collisions, the charge of the nucleus increases from +17 to +18, converting it into ^{37}Ar. However, the neutrino has removed a beta electron from the nucleus, so its mass is less than that of ^{37}Cl by about 0.000325 u, the difference seen above from adding a beta positron to it, making it 36.956252 u. The ^{37}Ar nucleus made from an electron neutrino interacting with ^{37}Cl is different from the one made from a proton or a deuteron interacting with ^{37}Cl.

Both versions of the ^{37}Ar nucleus can be converted back to ^{37}Cl by electron capture. In the ^{37}Ar nucleus formed from proton or deuteron collision with ^{37}Cl, classic electron capture occurs. A K-shell electron is pulled into the ^{37}Ar nucleus where it is split into a beta electron and an electron neutrino. The beta electron annihilates a beta positron in the ^{37}Ar nucleus, restoring it to the mass and charge of ^{37}Cl. The neutrino becomes a decay product of the process and carries off all the decay energy.

In the ^{37}Ar nucleus formed by neutrino interaction with ^{37}Cl, electron capture still occurs to transform the ^{37}Ar nucleus back into a ^{37}Cl nucleus. However, when the K-shell electron is pulled into the ^{37}Ar nucleus, instead of annihilating a beta positron to restore the ^{37}Cl nuclear configuration, its beta electron must become part of the nuclear configuration to replace the beta electron the nucleus lost in the ^{37}Cl interaction with the neutrino. The neutrino from the captured free electron is still released, but at a much lower energy since no beta electron – beta positron annihilation occurs.

In both cases, the electron capture causes the same compliment of Auger electrons to be emitted from the atom. The counters used at Homestake to count the ^{37}Ar recovered from experiments used the Auger electrons to see ^{37}Ar decays to ^{37}Cl. Therefore, the counter monitoring for the captures sees the same signal for the two types of decay. The counters cannot tell the difference between a ^{37}Ar made from a proton or deuterium collision with ^{37}Cl, or one made from ^{37}Cl interaction with a neutrino. But the two are different.

Because there are two different configurations of the ^{37}Ar nuclei, it is likely that they do not decay at the same rate. That is, their half-lives are probably different. If that is the case, then determining the number of neutrino interactions that occurred using the standard ^{37}Ar lifetime would give an incorrect value.

7.7 What Did the Chlorine Detector See?

All the analyses of the Homestake detector data was done using the lab ^{37}Ar half-life of about 35 days. There was no reason for those running the experiment and analyzing the data to believe that the half-life would be anything but 35 days. The Standard Model of Particle Physics does not indicate that the formation of a ^{37}Ar nucleus from ^{37}Cl could produce more than one configuration of the ^{37}Ar nucleus. The only way to know that the half-life was different would have been to analyze the ^{37}Ar decay data with the intent of determining its half-life.

The Homestake team did consider analyzing the decay data to determine if it indicated a 35-day half-life, but not to acknowledge that the ^{37}Ar was decaying at a different rate. Instead, the analysis was to confirm that they were analyzing ^{37}Ar. If the analysis had shown a different half-life, then the sample would have been considered either contaminated or just not ^{37}Ar. They would not have considered gas extracted from the detector with a half-life other than 35 days to be ^{37}Ar.

The count rates were so low for the ^{37}Ar samples extracted from the detector after ^{37}Cl exposure, that they were generally not useful for determining the ^{37}Ar half-life. However, there may be enough information in them to determine if the half-life of ^{37}Ar extracted from the detector was 35 day, or if it was greater than or less than that value. If the indication is that the half-life is less than the 35-day half-life of lab ^{37}Ar, it will be a mild validation of the ^{37}Ar formation models suggested above for proton/deuteron collision with the ^{37}Cl nucleus and neutrino interaction with the ^{37}Cl.

Table 7.3 gives a summary of the useful data taken from the Homestake detector from 1970 through 1975. There are 19 runs listed, with ^{37}Cl exposure times ranging from 33 to 216 days. After the ^{37}Ar gas was extracted from the detector, it would be counted for up to as many as seven times for a duration of about 35 days each time (one half-life).

Table 7.3: Homestake ^{37}Cl neutrino detector results from 1970 – 1975 [109]

Run no.	Exposure period		Counting periods (~ 35 days)							^{37}Ar atoms	
	Start date	Time (days)	1	2	3	4	5	6	7	Number in tank	Rate (per day)
18	4/12/70	216	5	3	2	1	1			30 ±13	0.60 ±0.26
19	11/14/70	112	10	5	4	4	3	3		29 ±14	0.63 ±0.30
20	3/6/71	103	3	1	3	0	1			8 ±10	0.19 ±0.22
21	6/17/71	107	1	3						5 ±16	0.11 ±0.36
22	10/2/71	72	2	2	3					1 ±12	0.03 ±0.26
23	12/13/71	80	4	0	1					-5 ±30	-0.12 ±0.75
24	3/2/72	77	2	1	1	0	0	1	0	10 ±3	0.25 ±0.23
27	7/7/72	121	11	6	4	3	1	2		55 ±18	1.19 ±0.40
28	11/5/72	82	4							16 ±16	0.40 ±0.40
29	1/26/73	78	6	1	6	2	3			10 ±15	0.25 ±0.38
30	4/14/73	139	1	5	1	3	4	1	2	8 ±11	0.17 ±0.23
31	8/31/73	104	1	0	1	1				-12 ±18	-0.28 ±0.40
32	12/13/73	43	1	1	1	1				-1 ±9	-0.05 ±0.33
33	1/25/74	152	3	2	2	1				15 ±13	0.31 ±0.27
35	7/1/74	33	0	4	2					2 ±14	0.08 ±0.58
36	8/3/74	194	4	6	3	2	0	0	1	27 ±12	0.68 ±0.30
37	2/13/75	121	6	3	3	0	0	1		33 ±13	0.81 ±0.32
38	6/14/75	102	7	3	4					23 ±14	0.53 ±0.32
39	9/24/75	121	6	2	3	2				17 ±12	0.33 ±0.26

The number of counts recorded in the first period after ^{37}Ar extraction are very low, with only runs 19 and 27 making it to 10 and 11, respectively. Most of them are less than six, and there appears to be no correlation between number of counts and exposure time.

Of the two runs with the highest counts, the source document indicated that run 27 may be contaminated. Therefore, run 19 will be used here to try to get a sense of what the half-life is of the neutrino induced ^{37}Ar. Its exposure time is t_{exp} = 112 days.

In run 19, there is a dramatic drop in counts from the first to the second counting periods. However, the number of counts in periods 3 through 7 are essentially constant. In fact, in most of the runs, the counts appear to level off above zero starting at period 3. This seems to suggest that, at least by counting period 3, there is a source producing ^{37}Ar within the counter to keep the counts constant.

The likely culprit is the ^{37}Cl gas that the ^{37}Ar decays back into. Likely muons from cosmic rays produce protons that interact with the ^{37}Cl from the ^{37}Ar decay, converting it back into ^{37}Ar. Clearly, something is making up the ^{37}Ar as it decays away beyond period 2.

The graph in Fig. 7.1 is a plot of the run 19 data broken into two segments. The first segment shows an exponential fit to the counts from periods 1 and 2. The second segment is a linear fit of the counts from periods 3, 4, 5, 6 and 7.

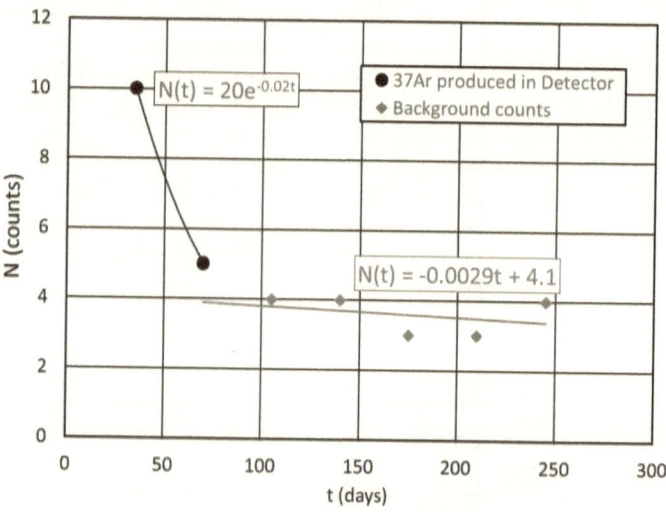

Fig. 7.1: Homestake ^{37}Cl neutrino detector run 19 data
The data from run 19 is broken into two segments. The first plots data from the first two counting periods and the second from the remaining five periods.

The first fit shows that initially the ^{37}Ar is decaying exponentially with a decay constant of λ = 0.02 s^{-1}, corresponding to a half-life of about 34 days and λ' = 0.022 s^{-1}. However, after 70 days, the counts stay constant at an average of four counts per counting period. This means that four ^{37}Ar atoms are being produced in the counter to re-place the four decaying during the counting period.

The graph in Fig. 7.2 plots the cumulative number of ^{37}Ar decays as a function of the time since the ^{37}Ar was extracted from the neu-trino detector. The fit shows that the ^{37}Ar decays at about 0.1 nuclei per day after the first counting period. This equates to about four de-cays each counting period. These counts are from ^{37}Ar made within the counter from ^{37}Cl built up in it. The count for the first counting period falls below the fit line. This likely indicates that, because the ^{37}Cl was building in the first period, less than four decays occurred.

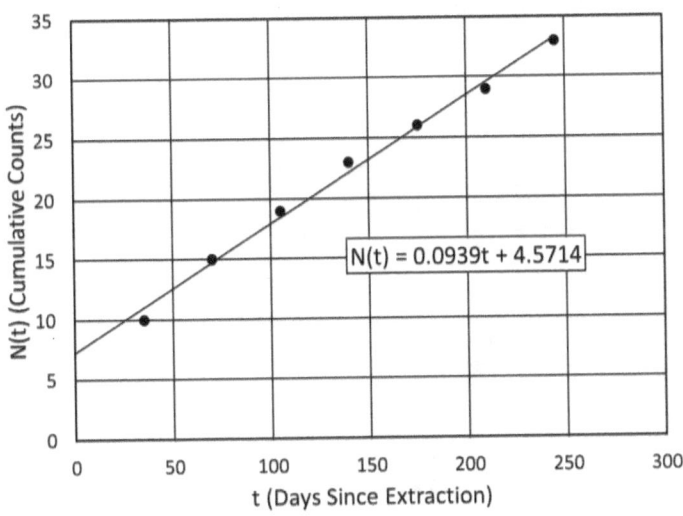

Fig. 7.2: Cumulative counts from Homestake detector run 19
The graph shows the total counts over time in run 19. The slope indicates that there is about 0.1 counts per day. This is generated by the decay of ^{37}Ar resulting from the buildup of new ^{37}Ar from conversion of the ^{37}Cl from the original ^{37}Ar decay.

Assume that only enough ^{37}Cl built up during the first counting period to produce two counts from the ^{37}Ar produced in the counter. In the count of 10 recorded in the first period, two of the counts came from the ^{37}Ar created in the counter, leaving eight counts from detec-tor ^{37}Ar.

In the second counting period, the ^{37}Cl in the counter built up so that three counts from that period were from new ^{37}Ar created from decay ^{37}Cl. That means there were only two counts produced by the original ^{37}Ar from the neutrino detector.

After the second counting period, there was no ^{37}Ar left from the original sample. All the counts registered were from new ^{37}Ar made from proton bombardment. This suggests that the original ^{37}Ar had experienced many more than two half-lives.

The graph in Fig. 7.3 shows the fit of the first two counting periods if the counts in the first period is eight and the counts in the second period is two. The decay constant for the original ^{37}Ar from the neutrino detector becomes $\lambda = 0.0400$ s^{-1}, which corresponds to a half-life of about 17 days and $\lambda' = 0.0404$ s^{-1}. The decay constant for a 35-day half-life is $\lambda = 0.0198$ s^{-1}, with $\lambda' = 0.0222$ s^{-1}. This shows that the neutrino-induced ^{37}Ar decays at a faster rate than the proton/deuteron – induced ^{37}Ar, indicating they are different.

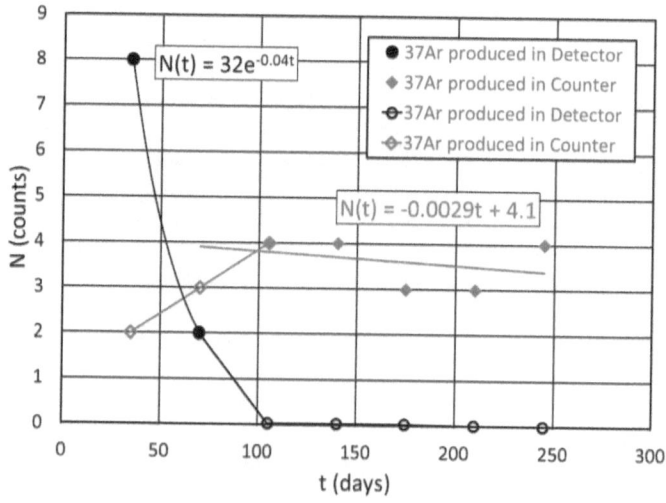

Fig. 7.3: Homestake ^{37}Cl neutrino detector run 19 with modified data
The count from the first counting period in run 19 is changed from 10 to 8 and the second period, from 5 to 2 to reflect the decay of ^{37}Ar resulting from the buildup of new ^{37}Ar from conversion of the ^{37}Cl from the original ^{37}Ar decay.

If this λ' is for the ^{37}Ar produced from ^{37}Cl interacting with neutrinos, then it is about 1.82 times that of the ^{37}Ar made from bombarding ^{37}Cl with protons or deuterons. That means the solar neutrino flux measured by the Homestake detector is actually about 1.82 times the value calculated using the lab ^{37}Ar decay constant. This would increase the flux value from 2.56 SNU to about 4.66 SNU, which is about 61% the calculated expected value of 7.6 SNU. This crude adjustment is reasonably consistent with the ^{37}Cl detector seeing 20 out of 37 neutrino interactions or 54% of the predicted neutrino flux.

This all suggests that there is nothing happening to the neutrinos coming from the Sun. The problem for the ^{37}Cl detector was in interpreting the measurements. The problem was subtle and aided in its elusiveness by the inaccuracy of the Standard Model depiction of the $^{37}Cl - v_e$ reaction, which produces the same ^{37}Ar configuration as the $^{37}Cl - p$ or $^{37}Cl - d$ reactions.

7.8 What Did the ES Reactions See?

The two ES reactions, H_2O and D_2O, claim to see a mixture of mostly electron neutrinos, but also muon and tau neutrinos, arriving at Earth from the Sun. Both SNO, seeing 2.39×10^6 cm^{-2}s^{-1}, and Super-Kamiokande, seeing 2.32×10^6 cm^{-2}s^{-1}, see the same flux. However, section 7.4 showed that the NC flux seen by SNO, 5.09×10^6 cm^{-2}s^{-1}, is all electron neutrinos. When combined with the CC flux, 1.76×10^6 cm^{-2}s^{-1}, always considered to be electron neutrinos, the total electron neutrino flux SNO sees is 6.85×10^6 cm^{-2}s^{-1}. This seems to indicate that SNO sees more than just ^8B neutrinos.

When the NC flux is considered the total neutrino flux from the Sun, the ES flux is about 45% of the total neutrino flux. However, if the total flux is actually 6.85×10^6 cm^{-2}s^{-1}, then the ES flux is only 35% of the total flux. Since all the neutrinos appear to be electron neutrinos, what is causing the ES reactions to record only 35% of them? It may be that not all the electrons in the water molecule can respond in a way that the detector can measure their interaction. In other words, the detector cannot see all the neutrino interactions.

In the H_2O and D_2O molecules, there are 10 orbital electrons available for the neutrino to interact with, eight from the oxygen atom and one from each of the hydrogen atoms. Two of the oxygen electrons are in the inner 1s orbital. If the neutrino interacts with one of them, they may not be able to escape the molecule with enough energy to produce the Cerenkov radiation.

Four more of the electrons are involved in bonding the two hydrogen atoms to the oxygen atom. Again, they may not be able to achieve Cerenkov velocities if hit by a neutrino. That leaves four electrons that are out in the open and free. These are the electrons that likely produce the Cerenkov radiation that the detectors see. This means that only four out of ten possible neutrino-electron interactions, 40%, may be detectable. This is consistent with the 35% the two ES reactions see.

7.9 Mystery Solved! – Maybe

Based on the conjecture and analyses of the previous sections, it seems the solar neutrino problem for the various neutrino detectors lie in not understanding and analyzing adequately how the neutrinos interact with the various detector materials. Consequently, the measured data were not properly interpreted, resulting in misidentification of the neutrino fluxes measured. This gave the appearance of the detectors not seeing a fraction of the neutrinos expected.

The charged current and neutral current reactions associated with the deuterons in SNO are not separate classes of interactions, but the same interaction occurring in different parts of the deuteron. There are three instances where the neutrino scatters off the deuteron, splitting into a proton and a neutron. There is one instance where the neutrino couples with a beta electron in the deuteron forming a free electron and splitting the deuteron into two protons. Consequently, the two reactions see two pieces of the same electron neutrino flux. Combined, they show all the electron neutrinos coming from the Sun and that there are no neutrino oscillations occurring.

The gallium detectors also reveal that there are no neutrino oscillations. While the gallium detectors see all the neutrinos coming from the Sun, it only registers about 55% of them. Like with the deuteron, the neutrinos only interact with beta electrons in the neutral nucleons in the gallium. They couple with a beta electron to form a free electron, converting the neutral nucleon into a positive one and the gallium into germanium. If the neutrino encounters a beta electron in a positive nucleon or a beta positron in either a positive or a neutral nucleon, it just scatters off the nucleus.

Unlike with the deuteron, where the two nucleons are held together by one bond and the neutrino interaction can split them, the nucleons of gallium are held in the nucleus by multiple bonds. A strike by a neutrino cannot dislodge one of them. When a neutrino scatters off a gallium nucleus, it produces no indication that it interacted with the nucleus. So, only 40 out of the 71 encounters the neutrinos have with the gallium, or 56%, are detectable.

The chlorine detector has the same neutrino interaction situation as the gallium. Only neutrino interactions with the 20 neutral nucleons within it register a hit in the detector. Therefore, only 20 out every 37, or 54%, of the neutrino interactions are seen by the detector. However, the chlorine detector appears to have another issue.

It seems that ^{37}Ar made by a neutrino interacting with ^{37}Cl is structurally different than ^{37}Ar made by bombarding ^{37}Cl with protons or deuterons. Consequently, the neutrino induced ^{37}Ar decays at a faster rate than the 35-day half-life of the other ^{37}Ar. As a result, when the detector data is processed using the 35-day half-life, the neutrino flux calculated is less than what the detector actually saw.

Processing of the chlorine data using a crude adjusted ^{37}Ar half-life showed the detector see about 60% of the predicted electron neutrino flux. This is more in line with the 54% expected to be seen because only 20 out of every 37 interactions will be detected.

The ES Cerenkov-water reactions in the Kamiokande and SNO detectors all claim to measure only about 45% of the solar neutrinos expected. Like with the other detection reactions, this appears to be due to only a fraction of the neutrino encounters with the electrons in the H_2O and D_2O molecules being detected.

Of the 10 orbital electrons in the molecules, two are in internal orbitals and four are involved in bonds. Neutrino interactions with these six electrons likely do not produce the Cerenkov radiation from the scattering needed for detection. Consequently, only four out of ten, or 40%, of the interactions are registered, consistent with the 45% measured by the detectors.

It seems there is nothing mysterious about neutrinos. The mystery appears to lie in how they interact with orbital electrons and components of nucleons. This is an area for which the Standard Model of Particle Physics does not offer any functional insight.

The root of the problem appears to be the oversimplified model of the neutrino–electron interaction that the Standard Model of Particle Physics offers. With the electron as a fundamental particle, it has no latitude to disassemble and reassemble during the interaction. It can only be converted between neutrino and electron via emission or absorption of a *W* boson.

8. Fundamental Particles

8.1 Some Leptons and Bosons Eliminated

The determination that the free electron, muon and tau particle are composites made of beta electrons and neutrinos eliminates three more particles in the Standard Model matrix of fundamental particles. The diagram in Fig. 8.1 shows the electron, muon and the tau crossed out of the last update to the matrix from Fig. 3.14.

Also, with the models of the baryons, mesons and leptons offered in the previous chapters; other, more traceable, mechanisms for particles to decay into other particles exist than the spontaneous appearance and subsequent decay of W or Z bosons. Consequently, they, too, can be crossed off the matrix as fundamental. Without W and Z, the Higgs boson, H, can also be removed.

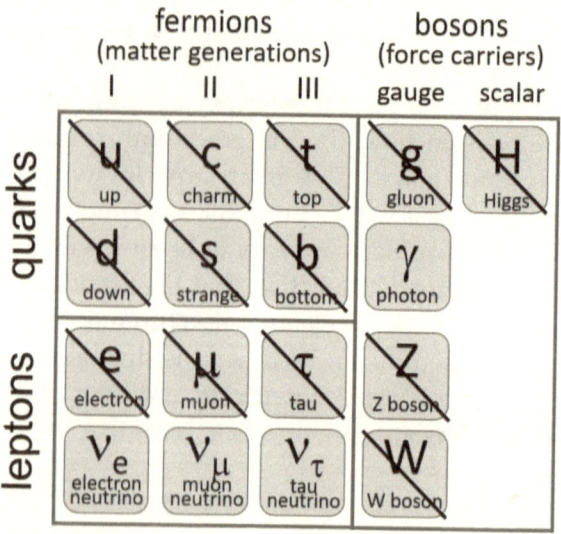

Fig. 8.1: Modified Standard Model
The Standard Model particle matrix showing the particles potentially eliminate by analyzing the structure of leptons. The electron, the muon and the tau are composites made of beta electrons and neutrinos. The new models of the leptons along with the models of the baryons and mesons eliminate the need for the virtual W and Z bosons to explain their decays. Without W and Z, the Higgs is no longer needed.

When baryons decay into mesons, they do so because the mesons are structural components of the decaying baryons, not because a virtual *W* or *Z* particle appears out of nothing and decays into those particles. Mesons decay into leptons because their structures are similar, and they can pair produce the neutrinos that convert them into the leptons.

By now, it seems clear that the fundamental components of nearly, if not, all the massive subatomic particles are the beta electron and its antiparticle, the beta positron. These particles, along with the three neutrinos and their antiparticles appear to form the baryons, the mesons and the leptons found in nature.

Though differing from the free electron by only an electron antineutrino, the mass of the beta electron is not readily apparent. However, because the mesons appear to be made of only beta electrons and beta positrons, a mass for the beta electron, relative to the free electron, can be determined.

8.2 Beta Particle Mass

The lightest of the mesons are the pions. The charged pions, π^+ and π^-, have a mass of 139.571 MeV or 273.132 free electron masses (fem), and the neutral pion, π^0, a mass of 134.977 MeV, or 264.143 fem. While the charged pions are slightly more massive than the muons; unlike muons, pions show no signs of consisting of anything other than beta electrons and beta positrons.

When the muon decays, it emits a muon neutrino, an electron (which contains an electron neutrino) and an electron antineutrino. The electron neutrino and antineutrino are the result of pair production during the decay. Because no muon antineutrino comes out of the decay, the muon neutrino that comes out of it is not the product of a pair production during the decay. It was in the muon before the decay occurred. The same is true of the antimuon decay.

When the charged pions decay, they essentially always decay into a positive muon and a muon neutrino, or a negative muon and a muon antineutrino. The positive muon or negative muon contains the sister neutrino to the neutrino that appears with it during the decay. The two neutrinos were formed via pair production during the decay. Consequently, the charged pions do not appear to contain anything but beta electrons and beta positrons before they decay.

The neutral pion provides an even clearer indication that it contains only beta electrons and beta positrons. Nearly all the π^0 decays result in only two γ photons being formed. The γ photons are the result of mass annihilation of the pairs of beta electrons and beta positrons contained within the π^0. Because the π^0 is neutral, it contains the same number of beta electrons as beta positrons; so, when the annihilations are done, usually no massive particles are left. No electrons, no muons and no neutrinos, only photons.

Less than 2% of the time a free electron and free positron are emitted along with a γ photon during a π^0 decay. In these cases, an electron neutrino-antineutrino pair formed during the mass annihilation. The two were captured by a beta electron and beta positron, respectively, before the betas could annihilate. This formed a free electron and a free positron that were emitted during the decay.

The masses of the pions in fem (free electron masses) are slightly greater than the apparent number of particles that form them. Because of this, the beta electrons and beta positrons inside the pions appear to be slightly more massive than a free electron.

In the case of the charged pions, the π^+ and π^-, with a mass of 273.132 fem, apparently contain 273 beta particles; 136 beta electrons and 137 beta positrons for the π^+ and 137 beta electrons and 136 beta positrons for the π^-. The 273 beta particles have a mass equivalent to 273.132 free electrons. That makes the mass of a beta particle 1.000484 times (273.132 ÷ 273) the mass of a free electron or, $m_\beta = 1.000484 m_e$.

The neutral pion, π^0, gives a slightly greater mass to the beta particles. With no net charge, the π^0 appears to be 264 beta particles, 132 beta electrons and 132 bet positrons. This makes the mass of the beta particles 264.143 ÷ 264 or 1.000542 times the mass of the free electron. Taking the average of the two, the mass of a beta electron inside a complex particle appears to be about 1.000513 times the mass of a free electron or, $m_\beta = 1.000513 m_e$.

The charged pions contain nine more beta particles than the neutral pion. With each having a mass of 1.000542 times the mass of a free electron, one might expect the mass of the charged pions to be about 9.004878 fem greater than that of the neutral pion or, 273.148 fem. However, the difference is only 4.5936 MeV or 8.9894 fem. This mass discrepancy likely indicates the beta particles experience a mass defect from being bound within the pion.

The graph in Fig. 8.2 plots the masses of the neutral and charged pions as a function of the apparent number of beta particles they contain. The linear fit of the two points is

$$m_\beta(n) = 0.99878n + 0.46567. \tag{8.1}$$

The fit gives the mass of betas in a particle, $m_\beta(n)$ as a function of the number of betas, n, in the particle. It shows that as the number of betas in a particle increases, the apparent mass of the individual betas $(m_\beta(n)/n)$, decreases. For example, the beta mass of the neutral pion containing 264 betas is 264.143 fem, making its betas 1.000542 times the mass of a free electron. However, adding just nine betas to it to create a charged pion containing 273 beta particles with a mass of 273.132 fem, suppresses that value to 1.000484.

The mass of the charged kaons, K+ and K- is 493.677 MeV or 966.100 free electron masses. At this mass, equation (8.1) predicts the kaons are made of 967 beta particles, producing a mass of 966.286 free electron masses. The neutral kaon, K0, has a mass of 497.611 MeV, or 973.798 free electron masses. The fit predicts the K0 is made of 974 beta particles with a mass of 973.277 free electron masses.

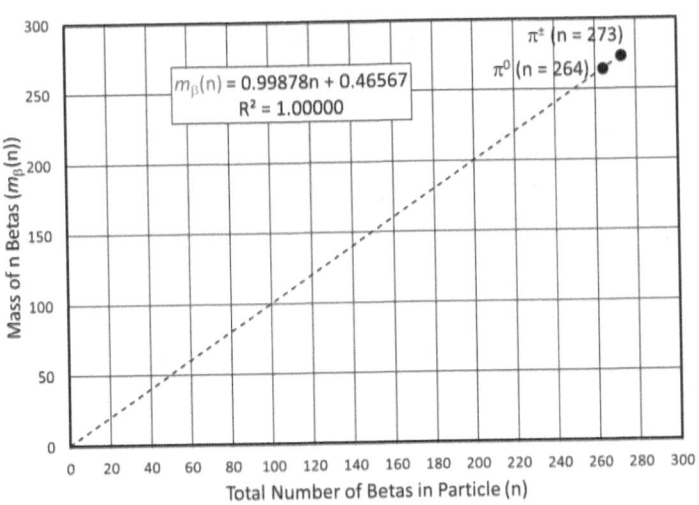

Fig. 8.2: Mass of Beta Particles in Pions
Graph shows the masses of the neutral and charged pions versus their apparent total number of beta electrons and beta positrons. The fit is a linear equation through the two points. The dashed line is the extrapolation of the fit back to the origin.

The charmed mesons D$^\pm$ and D^0 have masses of 1,869.3 MeV and 1,864.5 MeV, respectively; which makes their masses 3,658.121 and 3,648.728 free electron masses. Equation (8.1) predicts the charged D mesons are made of 3,661 beta particles, with a mass of 3658.997 free electron masses. It predicts the neutral D meson is made of 3,652 beta particles with a mass of 3,648.010 free electron masses.

The masses of the bottom mesons B$^\pm$ and B^0 are 10,330.724 and 10,331.507 free electron masses, respectively. Equation (8.1) predicts the charged B meson is made of 10,341 betas, giving it a mass of 10,330.847 free electron masses, and the neutral B is made of 10,342 beta particles, making its mass of 10,331.846 free electron masses.

8.3 Neutrino Masses

A free electron is a composite particle made of one beta electron. If equation (8.1) holds true down to n = 1, then it implies that the beta electron forming the free electron has a mass of 1.464 free electron masses. While this seems to indicate a problem with the extrapolation of the fit, there is another way to interpret it.

The mass of the free electron , m_e, is equal to the mass of one beta electron, $m_\beta(1)$, and the mass of the electron neutrino, m_{ve}, or

$$m_\beta(1) + m_{ve} = m_e.$$

If equation (8.1) is assumed to be valid down to n = 1, then the solution to this electron mass equation becomes

$$1.464 \text{ fem} + m_{ve} = 1 \text{ fem,}$$

or

$$m_{ve} = -0.464 \text{ fem.}$$

The mass of the electron neutrino becomes negative!

Based on the trend established by the masses of the pions, composite particles containing only beta electrons and beta positrons, the mass of the neutrino inside the free electron is less than zero. It is hard to say what that means physically, since all known physics seems to only deal in positive masses. However, if mass is a property of matter like charge, there is no reason why it cannot come in both positive and negative polarities.

This may sound far-fetched, but using this method yields a similar result for the mass of the muon neutrino. It, too, appears to be made of negative mass.

Recall the mass of a free muon is 105.658 MeV or 206.768 free electron masses. This apparently makes it 207 beta particles – 104 beta electrons and 103 beta positrons. With this, the mass of the free muon, m_μ, can be expressed as

$$m_\beta(207) + m_{v\mu} = m_\mu,$$

where $m_{v\mu}$ is the mass of the muon neutrino. Using equation (8.1), the muon mass equation becomes

$$207.213 \text{ fem} + m_{v\mu} = 206.768 \text{ fem},$$

or

$$m_{v\mu} = -0.445 \text{ fem}.$$

This seems to indicate that the mass of the muon neutrino is essentially the same as the mass of the electron neutrino. In fact, using the electron neutrino mass value for the muon neutrino mass in the free muon mass equation above gives a muon mass of

$$m_\mu = m_\beta(207) + m_{ve},$$
$$= 207.213 \text{ fem} + (-0.464) \text{ fem},$$
$$= 206.749 \text{ fem}.$$

On the other hand, the tau analysis gives a completely different result. With a mass of 1,776.860 MeV, the tau is 3,477.221 free electron masses. If the tau is made of 3,481 beta electrons and positrons, this makes the tau mass equation

$$m_\beta(3,481) + m_{v\tau} = m_\tau,$$

or

$$3,477.219 \text{ fem} + m_{v\tau} = 3,477.221 \text{ fem}.$$

Equation (8.1) predicts a tau made of 3,481 betas has a mass of 3,477.219 free electron masses, an almost exact match to the measured mass. This makes the mass of its neutrino

$$m_{v\tau} = 0.002 \text{ fem},$$

essentially zero.

Unlike with the electron and the muon – which implies the existence of a neutrino with negative mass – this result seems to imply the tau neutrino mass is near zero. This, along with the deviation of its anomalous magnetic moment from expected (sec. 4.6), strongly suggests the tau is not in the same family as the electron and the muon. Without a neutrino, the tau looks more like a meson than a lepton.

Nevertheless, strange D mesons, D_s mesons, are said to decay into tau leptons and their neutrinos just as they do into muons and their neutrinos. In doing so, assuming the tau contains the antiparticle of the neutrino released during the decay, clearly a neutrino-antineutrino pair was produced. Because of this decay, the tau does appear to have a neutrino and is a lepton like the electron and muon.

With the mass of the muon neutrino being essentially the same as the mass of the electron neutrino but dilated by the betas, it may be that the two are just one particle in different states. That is, there is only one neutrino whose mass changes as the number of betas it sees changes. This would align with the idea of neutrino oscillation, albeit in a slightly different sense than currently viewed.

Based on the electron and muon neutrino masses, assume that the negative mass of the neutrino dilates as the number of beta particles around it increases. If the dilation is linear, then the slope of the mass change from electron to muon is -0.019 fem/206 betas or -9.22 x 10^{-5} fem/beta particle. With a total of 3,481 beta particles, if the tau had the same neutrino, its mass would dilate by 0.321 fem to -0.143 fem from the electron neutrino mass. This would make the tau mass predicted by equation (8.1) 3,477.076 free electron masses, within 0.005% of the measured value.

Considering how the neutrinos for all three charged leptons seem to form, they could be all the same particle. That is, there is only one type of neutrino. The neutrino pairs for both the electron and the muon are modeled to form when a valence beta particle is engulfed in annihilation energy during a particle decay.

Apparently, the unit charge of the valance beta induces a neutrino-antineutrino pair from the annihilation energy. Unless the amount of energy available somehow influences the final mass of the neutrino spawned, a unit charge should always produce the same neutrino pair. In fact, once the requisite amount of annihilation energy is available, it seems that the neutrino-antineutrino pair would form. There would be no chance for the process to take advantage of any additional energy.

Therefore, there seems to be an argument for just one type of neutrino and its antiparticle in existence. The electron, muon and tau neutrinos appear to be the same. Only their circumstances, being in the presence of differing numbers of beta particles, causes them to appear as different particles.

One neutrino makes the world a lot simpler and is in line with the charge leptons being complex, not fundamental, particles. Since they are not basic particles, there is no need for them to have a unique partner neutrino to split from during weak interaction.

8.4 Impact of Negative Mass

So, how does the negative mass of the neutrino interact with the positive mass of the beta electrons and beta positrons? The typical belief is that negative mass should repel away from positive mass. However, since, based on the magnetic moment models (sec. 4.4 & 4.5), the beta particles appear to orbit the neutrino in both the electron and the muon; it could be that there is an attraction between the beta particles and the neutrino.

The apparent mass defect experienced by the neutrino as the number of beta particles increases also suggests some type of binding interaction. The proposed mass of the beta particle is $1.464m_e$ and the mass of the neutrino, $-0.464m_e$. The muon model proposed in section 4.5 puts their separation in both the muon and the electron at about 3×10^{-16} m. Assuming the law of universal gravitation holds for negative mass, the beta particle is pulling the neutrino toward it with an acceleration of $1.464g_e$, where $g_e = Gm_e/r^2$; while the neutrino is pushing the beta particle away from it with an acceleration of $0.464g_e$.

The net acceleration should cause the beta particle and the neutrino to accelerate toward each other at a rate of g_e, which turns out to be about 2×10^{-25} m/s². This does not produce a force strong enough to counter the centrifugal force produced on the beta particle by its high-frequency orbit around the neutrino.

If there is some yet unknown source of attraction, it does not appear to be associated with charge, since in the muon, the neutrino appears to attract both beta electrons and beta positrons. Although, in both the electron and the muon, the neutrino seems to form the negative version of the particle (e^- and μ^-), whereas the antineutrino, the positive version (e^+ and μ^+).

The same appears to be true regarding the matter making up the beta particles. Assume the matter making up the beta positron is somehow different from the matter making up the beta electron, since it is antimatter. Again, the muon neutrino appears to be indiscriminate of the type of matter of the particle, since it accommodates both beta positrons and beta electrons in forming the muon.

There appears to be some property other than mass and charge that particles have that connects the neutrino to the beta particles in leptons. The property probably manifests itself as a state in beta electrons and beta positrons, and the anti- to that state in neutrinos. The property apparently does not cause an interaction between like instances of itself, such as with two beta electrons, or a beta electron and a beta positron, or two neutrinos, or a neutrino and an antineutrino. However, a particle in one state interacts strongly with a particle in the anti-state, if close enough to it, such as a beta electron and a neutrino or a beta positron and an antineutrino.

8.5 What's Left?

The chart in Fig. 8.3 shows the additional Standard Model particles eliminated as a result of the neutrino mass analysis. The analysis revealed there is not an electron neutrino, a muon neutrino and a tau neutrino, but likely a single "universal" neutrino. That neutrino and its antineutrino appear to serve in the configuration of the electron, the muon and the tau leptons.

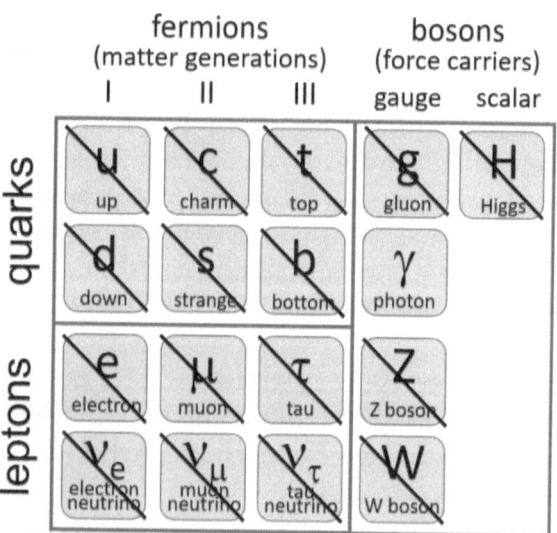

Fig. 8.3: Modified Standard Model
The Standard Model particle matrix showing the particles, in addition to those shown in Fig. 8.1, potentially eliminate by analyzing the neutrino mass. There appears to be no tau neutrino and no separate electron and muon neutrinos. This eliminates all the particles in the matrix except the photon.

Consequently, Fig. 8.4 shows that the particles shown as fermions in the Standard Model are replaced by the beta particle, β, the neutrino, v, and their antiparticles. This reduces the matter-generating fermions from the 12 shown in the Standard Model table, to two – the β and the v. The only boson left is the photon.

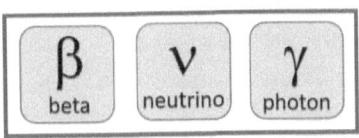

Fig. 8.4: Beta Model
The Beta Model particle matrix showing the particles revealed as fundamental after electron-proton deep inelastic scattering and lepton analyses. They include the beta electron, a universal neutrino, the photon and the appropriate antiparticles.

Some 17 particles required by the Standard Model to model matter and interactions have been reduced to three. The only survivor from the Standard Model – the photon. All the baryons, mesons and charged leptons can be made of combinations of the β^+, β^-, v and $\bar{v}$ particles.

The free electron is a β^- in orbit around a v and the free positron is a β^+ orbiting a $\bar{v}$. The proton is made of eight pions whose combined polarity is +1. The charged pions are made of a cluster of 33 β^-– β^+ pairs plus an additional β^- for the π^- or β^+ for the π^+, surrounded by a shell of 103 β^-– β^+ pairs. The neutral pion, π^0, has 32 β^-– β^+ pairs clustered inside the shell.

9. Complex Particles

After carefully analyzing and evaluating the behaviors of the particles deemed fundamental in the Standard Model, all but the photon have been shown to be either complex particles, made of collections of particles, or to likely not exist at all. Out of these reviews, two particles have emerged that appear to truly be fundamental, the beta particle and the neutrino.

The beta particle comes in matter form as a beta electron and in antimatter form as the beta positron. It is akin to the electron, a particle that was thought to be fundamental in the Standard Model; but was shown to be a composite of the beta particle and the other fundamental particle, the neutrino.

Like the beta particle, the neutrino has similar counterparts in the Standard Model. However, the Standard Model hosts three types of neutrinos as fundamental: the electron, muon and tau neutrinos. Deeper analysis discussed in chapter 8 showed that they seem to all be the same particle.

While new descriptions of the leptons and hadrons from the Standard Model have been offered in earlier chapters, there have been no discussion of why or how these particles configure themselves the way they do. The following analyses and evaluations attempt to uncover and provide some rhyme and reason as to why the complex particles are the way they are.

9.1 Subnuclear Structure

As discussed in previous chapters, complex particles such as baryons and mesons appear to be made of collections of pions. Pions decay into muons, and muons into electrons. Pions and muons appear to share a structural feature, a subnuclear shell of beta positrons and electrons, which the muon inherits from the pion decay.

The muon decays into an electron, during which all but one of the beta electrons or positrons in the shell get annihilated. However, the remaining beta particle appears to retain the radius of the muon shell. That the orbital radius of the beta particles in the electron, muon and pion are essentially the same suggests that the radius of the shell is a fundamental configuration. Beta electrons and positrons naturally conform to it like electrons to the Bohr radius in atoms.

The four diagrams in Fig. 9.1 show the four complex particles discussed here, the μ-kaon, the pion, the muon and the electron. They all are shown with the subnuclear shell made of 35 orbitals. In the kaon, pion and muon, 34 of the orbitals contain six beta particles, three beta positrons and three beta electrons.

The 35th orbital of the kaon and the pion contain only two beta particles, a beta electron and a beta positron. The valence beta particles at the center of the charged versions of these particles migrates to the partially filled orbital when they decay into a muon, giving it three betas and making it a charged orbital.

The muon has the three beta particles in its 35th orbital, a beta electron and beta positron plus the valence beta, thanks to the decay of the kaon or pion. The electron is shown with all 35 orbitals, but only one beta particle in one of its orbitals. The other orbitals are there to show that it shares the structure of the other three particles.

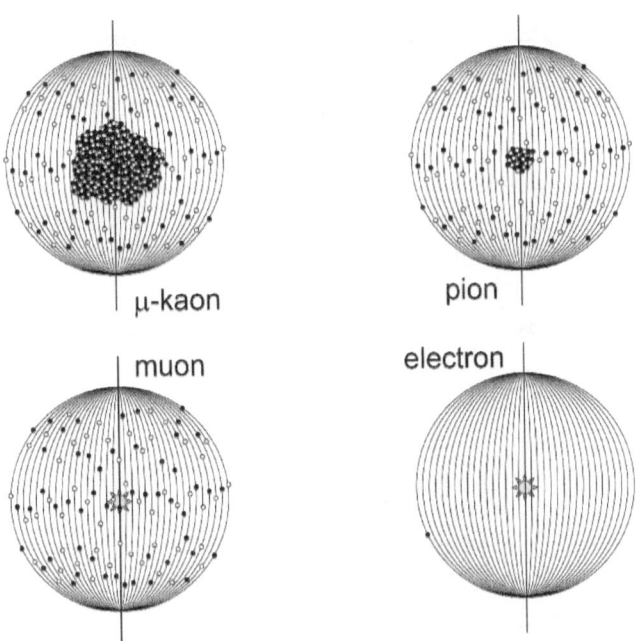

Fig. 9.1: Subnuclear Shell of Complex Particles

The diagrams show that K_μ kaon, the pion, the muon and the electron all share a common shell structure that holds three beta electrons and three beta positrons in all but one of the orbitals. The electron is made of the shell but has only one beta particle in one orbital in it.

119

The radius of the orbitals that make up the shell is estimated at 3.2×10^{-16} m, about 22.5 times the radius of a beta particle. There are 35 meridian orbitals that cross each other at the poles of the shell and each orbital can hold up to six beta particles, three beta electrons and three beta positrons.

It was suggested in section 8.4 that the central neutrino in the free electron and muon may somehow be attracting the beta electrons and positrons, causing them to orbit the neutrino in the lepton. However, there is no neutrino at the center of the pion or kaon, yet the electrons and positrons still travel the same orbits. Therefore, falling into orbitals of that radius appears to be an inherent behavior of the free beta particles and not caused by a central attractor.

When all the orbitals in the subnuclear shell are filled, beta particles can apparently overflow into a cluster at the center of the volume enclosed by the shell. This is how the pions, some kaons and possibly some D mesons form.

A cluster of 33 pairs of beta electrons and positrons and an unpaired valence beta particle inside a filled shell less three betas make a charged pion. A cluster of 32 beta electron-positron pairs in the shell form a neutral pion. The same shell encapsulating 380 β^-– β^+ pairs and a valence β forms the μ version of the kaon, K_μ, discussed in section 5.2. Why 33 pairs or 32 pairs or 380 pairs? There is no apparent explanation for now except that is just the way it is. Eventually, the reasons will probably reveal themselves.

Nevertheless, there appears to be a subnuclear configuration that beta particles naturally fall into when they are created. Therefore, if an event creates hundreds of pairs of beta electrons and positrons, initially those particles would naturally assemble themselves into pions and kaons. Some of those particles would bind together to momentarily form more complex mesons and baryons.

Given their mean lifetimes ($<10^{-8}$ seconds), the baryons would quickly decay down until they became stable protons. Meanwhile, the mesons would decay into pions and kaons. These, along with the pions and kaons that did not become components of mesons or baryons, would also almost instantly decay down into muons, which would decay further into stable electrons in less than one second.

Now, in a matter of seconds after the formation of the beta particles, all the ingredients for making hydrogen atoms, the primordial material of the universe, are formed.

9.2 Beta Particle Formation

When the 103 pairs of beta electrons and beta positrons in the shell of a muon annihilate during its decay, the charge field of the valence beta particle there produces a neutrino-antineutrino pair from the energy the annihilations produce. The valence beta then captures the neutrino, if it is an electron, or the antineutrino, if it is a positron, to form a free electron or positron.

This process resembles the pair production of an electron and a positron from a photon, shown in Fig. 9.2. When a photon of suffi-cient energy (>1.022 MeV) enters the charge field of a nucleus, it splits into a free electron and a free positron. Since the resulting particles contain neutrinos, apparently the combination of the charge field of the nucleus and the energy of the photon creates a neutrino-antineu-trino pair, just as in muon decay.

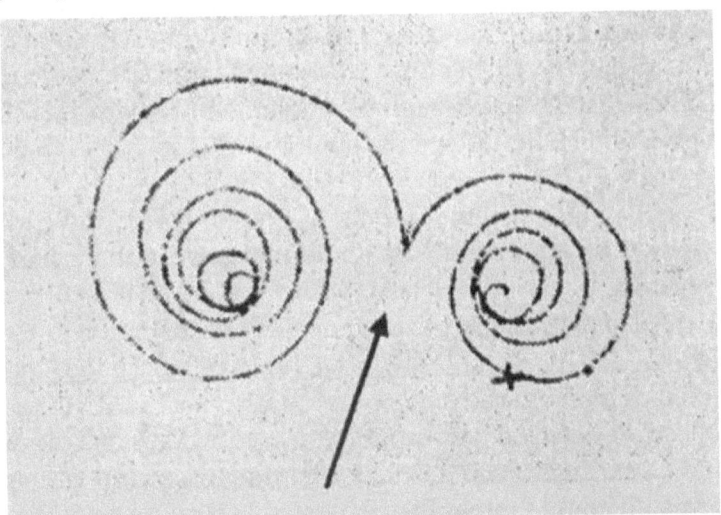

Fig. 9.2: Electron-Positron Pair Productions from Photon
The diagram shows a bubble chamber photograph of an unseen gamma photon split-ting into a positron (left spiral) and electron (right spiral) near an atomic nucleus. The path of the gamma photon is along the arrow.

However, in a reversal of the process in muon decay, the presence of the newly-formed neutrino-antineutrino pair causes the photon to split into what are apparently its components – a beta electron and beta positron. Once the split occurs, the neutrino pairs with the beta electron to form a free electron, and the antineutrino with the beta positron to form a free positron.

In low-energy annihilations of free electrons and free positrons, photons form, usually two to conserve the momentum of the two annihilating particles before the collision. Apparently, the neutrino from the electron and antineutrino from the positron are liberated, but because they are chargeless, go undetected.

In high-energy electron-positron annihilations, large complex particles can form if the collision provides enough energy to produce the required mass. For example, high-energy electron-positron annihilations can produce muon-antimuon pairs containing a total of 207 beta electrons, 207 beta positrons, a neutrino and an antineutrino, or tau-antitau pairs with over 3400 beta electrons and 3400 beta positrons along with the neutrino and antineutrino. Something about the collision transforms the energy associated with it into beta electrons and beta positrons.

In the Standard Model, the collision of the electron and the positron supposedly forms a Z boson, which decays into the lepton-antilepton pair. However, the Z decay assumes the leptons are fundamental particles. That is, a muon is a single particle, or a free electron is a single particle. It has been shown in earlier chapters that that is not the case.

To form a muon-antimuon pair, a Z boson would have to decay into over 400 beta particles. To decay into a tau-antitau pair, the Z must produce nearly 7,000 beta particles. Even in the Standard Model, the Z boson appears to be capable of producing only two particles at a time. These points are moot, however, since the Z has been eliminated as a real particle.

In electron-positron pair production, a gamma photon appears to split into a beta electron and beta positron as a neutrino-antineutrino pair forms. It could be that the photon is a coupling or mixture of a beta electron and a beta positron; and bringing a neutrino and an antineutrino together creates a field that causes the betas to repel or separate from each other.

Or, not necessarily a mixture of a beta electron and a beta positron, but of matter and antimatter. The field created by the neutrino-antineutrino pair separates the matter and antimatter in the photon and causes each of them to break up into fundamental packets of 0.51 MeV, which are the beta electrons and beta positrons. Any leftover matter and antimatter less than 0.51 MeV each recombine back into energy that gets absorbed by the particles formed.

In other words, during electron-positron annihilation, when the neutrino in the electron is brought together with the antineutrino in the positron, a field is produced that splits all the energy in the vicinity into its equivalent of beta electron – beta positron pairs.

Once formed, some of the beta particles gather together in orbits in the subnuclear shells discussed in the previous section. This is a fundamental configuration the beta electrons and beta positrons tend to assume. The neutrino and antineutrino from the electron and positron that initiated the annihilation are captured by a negative and positive shell, respectively, to form a muon and an antimuon.

If the energy is high enough such that more than two subnuclear shells are created, some beta particles gather in clusters at the centers of the volumes the extra shells enclose. These forms pions and K_μ kaons. And, if the energy is extremely high, it may be that additional beta particle shells encloses the muon and antimuon forming a tau and an anti-tau pair. The extremely high-energy collisions can also create heavy mesons.

9.3 Photon Structure

Electron-positron pair production from a photon, along with pair annihilation of a beta electron and beta positron, suggest that the photon is some combination of a beta electron and beta positron. During the decay of complex particles such as pions, beta electrons and beta positrons come together and transform into energy. When other particles, such as valence betas, are left over, they absorb that energy and convert it into motion. However, when the beta electrons and beta positrons are completely used up in annihilations, such as in the decay of neutral pions, the energy produced manifests itself as photons. The beta electrons and beta positrons become photons.

Unlike the beta electrons and the beta positrons, the photons are said to have no mass. However, what does "having no mass" really mean? Mass gives bodies two things: energy ($E = mc^2$) and momentum ($p = mc$). The energy and momentum a body has are proportional to its mass. Bodies with more mass under similar conditions possess more momentum and energy than bodies with less mass. However, photons also possess both energy ($E = h\nu$) and momentum ($p = h/\lambda$) but are considered massless. So, are they really massless, or massive, but lacking some other attribute that makes them different from the so-called "massive" particles?

What is different about the essence of a photon compared to a massive particle like a beta electron or beta positron? One thing that photons can do that massive particles cannot is occupy a space but not use it up. Massive bodies, like beta electrons and beta positrons can also occupy the space, but they use up the space they occupy.

Once occupied, the space a massive body takes up cannot be occupied by another massive body. But two photons can be in the same space and the massless photon can occupy the same space as a massive body through its absorption by the body. Therefore, there is some attribute or property that massive bodies have that prevents two of them from sharing the same space. It seems the likely culprit is electric charge.

Because photons have energy and momentum implies that they also have mass. The mass of a photon is its energy divided by the speed of light squared ($m_{photon} = E/c^2 = h\nu/c^2$) or its momentum divided by the speed of light ($m_{photon} = p/c = h/\lambda c$). So, technically, photons do have mass. What they do not have is charge. But particles such as beta electrons and beta positrons do have charge.

Along with its mass, the beta electron's unit negative charge makes it matter. The beta positron's unit positive charge combined with its mass makes it antimatter. With no charge, the photon, even with mass, is neither matter nor antimatter – call it *submatter*.

It appears that bodies made of matter or antimatter like beta electrons and beta positrons, are impenetrable to other matter/antimatter bodies. However, submatter bodies, with mass but no charge, like photons, are invisible to other bodies. A beta electron with a unit negative charge uses up the space it occupies so that no other matter particle can share the space. The same is true of the beta positron.

The beta electron and the beta positron both carry the same amount of mass, but their charges make them respond differently to their surroundings. When a beta electron encounters another negatively charged beta electron, the like charges will not allow the two to share space. The same is true when a beta positron encounters another positively charged beta positron.

However, when a beta electron encounters a beta positron, the two being oppositely charged, cancel each other's charges, but leaves the masses intact. Because the combined masses have no charge associated with them, they are now submatter and do not use up the space they occupy, they share it. Together they become a photon.

The charge that mass emanates is apparently associated with the mass emanating it. In other words, mass that emanates positive charge is different from mass that emanates negative charge. The type of mass determines whether a particle is matter or antimatter. Because the types of matter are different, in the presence of an electric field, the two will tend to separate.

As discussed in the previous section, in the presence of an electric field, energy can spawn the formation of a neutrino-antineutrino pair. If the energy of a photon is at least 1.022 MeV, a neutrino-anti-neutrino pair forms and splits the photon into its matter and antimatter components, the beta electron and beta positron. The neutrino has an affinity for the matter and the antineutrino, the antimatter.

That is why photons with enough energy will divide into a beta electron and a beta positron in the presence of a strong electric field such as a nucleus. However, the photon must contain enough energy (submatter) to make at least one beta electron and one beta positron for the submatter to separate. Since the neutrino and antineutrino are doing the splitting, this suggests that the photon must have at least 1.022 MeV of energy to spawn the neutrino-antineutrino pair.

A 1.022 MeV photon will split into a beta electron and a beta positron in the presence of a neutrino-antineutrino pair. However, a 2.044 MeV photon will not split into two beta electrons and two beta positrons. It splits into one of each and divides the remaining energy between the two as kinetic energy. This is likely because the photon can produce only one neutrino-antineutrino pair, regardless of its energy content. It signals that the photon is made of just one pair of beta particles. As a result, only one neutrino-antineutrino pair can be formed. Any energy the photon has over 1.022 MeV is kinetic energy, not mass energy. This suggests that all photons are made of only one beta electron and one beta positron coupled together.

Photons with less than 1.022 MeV of energy are apparently still made of a coupled beta electron – beta positron pair, but the pair is somehow bound together, causing mass defect within the photon. The binding energy reduces the amount of matter and antimatter within the photon so that, in an electric field, it cannot produce the neutrino-antineutrino pair to separate the matter and antimatter. Consequently, the particle remains a photon. What this all says is that, the photon, while appearing to be a fundamental particle, is, in fact, a complex particle made of a beta electron and a beta positron.

10. Hitting the Mark!

Determining that all photons are made of a beta electron and a beta positron coupled together has reduce the number of apparent fundamental particles to four: the beta electron, the beta positron, the neutrino and the antineutrino.

The beta electron and the beta positron are the fundamental packets matter and antimatter can come in. They are (anti)matter quanta, the fundamental units of (anti)matter. No other packaging of matter and antimatter in fractional or multiple quantities of this size occurs in nature. All (anti)matter objects larger than beta particles are made of collections of these (anti)matter units.

10.1 A Different Kind of Stuff

The analyses from section 8.3 showed that the neutrino and the antineutrino appear to be made of negative mass. Negative mass could be thought of as *antimass*, not to be confused with antimatter. Matter is positive mass that emanates negative charge, whereas antimatter is positive mass that gives off positive charge. Antimass is negative mass.

Neutrinos are thought to be chargeless because they do not respond to magnetic fields in the matter world as charged matter particles do. However, the fact that they do seem to be drawn to specific charge types of matter particles hints at veiled charges. Antimass likely manifests itself as mass does in that some antimass gives off negative charge and some emanates positive charge. Stuff made of positive mass is called matter. To give it a name, call stuff made of negative mass, *manner* (a play on the word manna).

Assume oppositely charged matter and manner still attract even though the masses are different. Since the neutrino appears to couple with the negatively charged beta electron to form a free electron, it is probably made of negative mass emanating positive charge.

To retain the convention already in place for neutrinos, unlike matter, call antimass that emanates positive charge *manner*, and antimass that gives off negative charge, *antimanner*. This means that the antineutrino is made of negative mass giving off negative charge or antimanner, consequently attracting the positively charged beta positron to form a free positron.

10.2 A Four-Particle Universe

It appears the universe is made of four particles, the beta, the neutrino and their antiparticles. This sets up an interesting way to characterize the fundamental universe. As shown in Fig. 10.1, if the abscissa of a two-by-two matrix is charge and the ordinate, mass, the quadrants represent the four types of stuff that make up the universe: matter, antimatter, manner and antimanner.

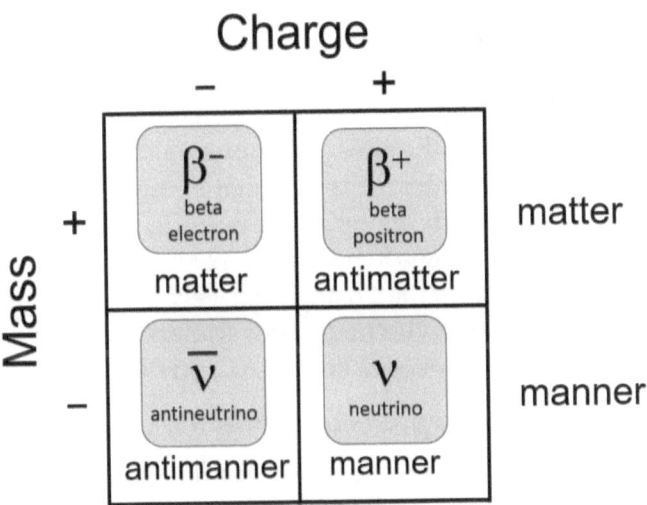

Fig. 10.1: The Four Fundamental Particles
The diagram shows a matrix of the four types of stuff that make up the universe. The four types of fundamental particles, the beta electron (β^-), the beta positron (β^+), the neutrino (v) and the antineutrino ($\bar{v}$) fall in quadrants of a matrix with charge as its abscissa and mass as its ordinate.

The perception is that matter comprises the bulk of the visible stuff in the universe. However, it has been shown here that pions, the components of protons, are made of nearly equal numbers of beta electrons and beta positrons, matter and antimatter. The mismatch in numbers of matter and antimatter particles in the proton is one more antimatter particle than matter particle.

The imbalance is compensated for by the electron, which is made of one matter particle, the beta electron. Consequently, since a neutral atom has an equal number of protons and electrons, it is made of an equal amount of matter and antimatter. In that sense, the universe is balanced.

As for the manner particles, neutrinos and antineutrinos, it is widely acknowledged that they fill the universe and do not routinely interact with matter. However, a neutrino is also a component of every free electron in existence. This makes them equal in numbers to protons and electrons in neutral atoms.

When a beta electron escapes a complex particle during beta decay, to become a free electron, it creates a neutrino-antineutrino pair. The beta electron captures the neutrino to become a free electron, while the antineutrino goes free. Since free positrons are not commonly observed in nature, the free antineutrino from the pair must be out there, unmatched by a free neutrino.

Supposedly, there are countless neutrinos whizzing through everything on Earth all the time and not even being slowed down by anything they encounter. It is said that manner (neutrinos) does not normally interact with matter. This is evidenced by the observation that lightyears of matter do not stop most neutrinos. However, there are neutrino detectors that, using relatively finite amounts of materials, allegedly interact with neutrinos. Materials such as ^{37}Cl and ^{71}Ga do react with neutrinos becoming ^{37}Ar and ^{71}Ge, respectively.

10.3 Quarks and Leptons and Bosons, Oh My!

The Standard Model's menu of fundamental particles is apparently not fundamental and, in some cases, the particles are not even real. The six quarks and their antiparticles do not appear to exist, even though the Standard Model has woven them into the fabric of the universe. The "scheme" using quarks and gluons can be made to model particles such as baryons and mesons, but they are not supported by deep inelastic scattering experiments. When the whole picture is viewed rather than increments, there is no basis for a three-particle proton and therefore, no basis for quarks. Baryons appear to be made of mesons and mesons of beta electrons and positrons.

The leptons, designated fundamental by the Standard Model, clearly show themselves as composites through their decay into lesser particles. The tau decays into pions, muons and electrons, and looks more like a meson than a lepton. The muon decays into an electron and two neutrinos. Not the signature of a fundamental particle. The electron, though stable, reveals its complexity through its magnetic moment that is slightly greater than it would be if the electron were a single particle.

The force-carrying particles – the bosons – also appear to be complex particles, when they appear to exist. The gluon, which is a pair of quarks, cannot exist if quarks do not exist. Though the gluons are thought to hold nucleons together in the nucleus of atoms, other mechanisms such as particle sharing, and even electromagnetic forces, could also do the job. Consequently, gluons do not exist.

The photon was shown earlier to be a complex particle consisting of a beta electron and beta positron coupled together. Although it does exist, it is not fundamental. That is why it can split into an electron and a positron as a result of pair production.

The functions that the W and Z bosons serve in the Standard Model, emission from complex particles and decaying into lesser particles, can easily be explained with models made of electrons and positrons. There is no need to have these particles appear out of nowhere and decay into the particles needed to explain the observed transformation. Radioactive decays are caused by particle annihilations not weak nuclear forces.

Finally, if the W and Z are not force carriers, then there is no need for the Higgs boson. The Higgs was invented so that the mass of a particle is not an inherent property of the particle, but the result of the particle interacting with the Higgs field. This allowed the W and Z to interact with the Higgs field to have mass without actually possessing mass. Then, the symmetry with the photon and gluon needed to combine the electromagnetic and weak forces into the electroweak force is not broken. There is no weak force causing radioactive decay; therefore, the electroweak force is not real. It is another "scheme," another modeling device set up by the Standard Model to mathematically simplify observed natural phenomena.

The bottom line is that the effort to make the patchwork of theories and ideas that form the Standard Model work are what causes its failure. When a theory or idea was proposed, the impetus from that point on was to find a way to interpret nature so that the idea or theory was validated. This was demonstrated when the particles found inside the proton were willed into being quarks, and when nearly 50 years and billions of dollars were relentlessly dedicated to producing a result that could be called the discovery of the Higgs boson. The result was a universe containing at least 37 so-called fundamental particles (12 quarks, 12 leptons, 8 gluons, 2 W bosons, a Z boson, a Higgs boson and a photon). It seems the mark was missed.

References

1 Emmanouil Magiorkinis, *et al.*, "The philosophic and biological views of the atomic philoso-phers, Leucippus and Democritus," *Hellenic Journal of Nuclear Medicine*, May – August 2010, http://www.nuclmed.gr/wp/wp-content/uploads/2017/03/9-1.pdf, and in section 2 of Sylvia Berryman , "Ancient Atomism", *The Stanford Encyclopedia of Philosophy* (Winter 2016 Edition), Edward N. Zalta (ed.), https://plato.stanford.edu/archives/win2016/entries/at-omism-ancient/.

2 Pieter Sjoerd Hasper, "Aristotle's Diagnosis of Atomism" *Aperion*, 39 (2006): 121.

3 R. Boyle, *A Continuation of New Experiments, Physico-Mechanical, Touching the Spring and Weight of the Air, and their Effects* (Oxford: H. Hall, 1669), http://tei.it.ox.ac.uk/tcp/Texts-HTML/free/A29/A29007.html .

4 John Dalton, *A New System of Chemical Philosophy*. (London: R. Bickerstaff, Strand, 1808), http://digbib.ubka.uni-karlsruhe.de/volltexte/digital/1/277.pdf.

5 D. Mendeleev, Osnovy Khimii (The Principles of Chemistry), 5th ed., Ch. 15, 448-472.

6 J.J. Thompson, "Cathode Ray," *Philosophical Magazine* 44 (1897): 293.

7 Ernest Rutherford, "The Scattering of α and β Particles by Matter and the Structure of the Atom," *Philosophical Magazine* 6 (1911): 21.

8 Ernest Rutherford, "Collisions of Alpha Particles with Light Atoms. IV. An Anomalous Effect in Nitrogen," *Philosophical Magazine* 6 (1919): 37; Oliver Lodge, "Letter to the Editor," *Nature* 106 2663 (1920): 467; Editors, "Physics at the British Association," *Nature*, 106, 2663 (1920): 357.

9 James Chadwick, "The Existence of the Neutron," *Proc. R. Soc. London A*, 136 (1932): 692.

10 James Clerk Maxwell, "A dynamical theory of the electromagnetic field," *Phil. Trans. R. Soc.* 155 VIII (1865): 459. http://doi.org/10.1098/rstl.1865.0008

11 Albert Einstein, "Grundlage der allgemeinen Relativitätstheorie (The Foundation of the Gen-eral Theory of Relativity)," *Annalen der Physik (ser. 4)* 49 (1916): 769.

12 Carl Anderson, "The Positive Electron," *Phys. Rev.* 43 (1933): 491.

13 Seth Neddermeyer and Carl D. Anderson, "Note on the Nature of Cosmic-Ray Particles," *Phys. Rev. Lett.* 51 (1937): 884.

14 Cecil Powell, *et al.*, "Observations on the Tracks of Slow Mesons in Photographic Emulsions," *Nature* 160 (1947): 453.

15 George Rochester and Clifford Butler, "Evidence for the existence of new unstable elementary particles," *Nature* 160 (1947): 855.

16 Clyde Cowan and Fred Reines, "Detection of the Free Neutrino: A Confirmation," *Science* 124 (1956): 3212.

17 Leon Lederman, *et al.*, "Observation of High-Energy Neutrino Reactions and the Existence of Two Kinds of Neutrinos," *Phys. Rev. Lett.* 9 (1962): 36.

18 M. Gell-Mann, "A Schematic Model of Baryons and Mesons," *Phys. Lett.* 8 (1964): 214.

19 G. Zweig, "An SU(3) model for strong interaction symmetry and its breaking II," CERN-8419-TH-412, (1964).

20 Arie Bodek, "The Structure of the Nucleon, Three Decades of Investigation (1967-2004)," *Nuc. Phys. B* - Proceedings Supplements, 139 (2005): 165.

21 Steven Weinberg, *et al.*, "A Model of Leptons," *Phys. Rev. Lett.* 19 (1967): 1264.

References

[22] A. Salam, in Elementary Particle Theory, Almqvistand and Wiskell 367, 1968.

[23] Tian Yu Cao, *Conceptual developments of 20th century field theories*, (Cambridge, Cambridge University Press, 1998).

[24] Abraham Pais and Samuel Treiman, "Neutral heavy leptons as a source for dimuon events: a criterion," *Phys. Rev. Lett.* 35 (1975): 1206.

[25] M. Tanabashi *et al.*, (Particle Data Group), "Review of Particle Physics," *Phys. Rev. D* 98 030001 (2018): 40. http://pdg.lbl.gov/2019/listings/contents_listings.html.

[26] Tanabashi, Particle Physics, 36.

[27] Tanabashi, Particle Physics, 33.

[28] H. Georgi and S.L. Glashow, (1974). "Unity of All Elementary Particle Forces," *Phys. Rev. Lett.* 32 (1974): 438.

[29] B.P. Abbott BP, et al. (LIGO Scientific Collaboration & Virgo Collaboration), "Observation of Gravitational Waves from a Binary Black Hole Merger". *Phys. Rev. Lett.* 116 (6) (2016): 061102.; B.P. Abbott BP, et al. (LIGO Scientific Collaboration & Virgo Collaboration), "GW170817: Observation of Gravitational Waves from a Binary Neutron Star Inspiral,". *Phys. Rev. Lett.* 119 (16) (2017): 161101.

[30] S. Chatrchyan, *et al.*, (CMS Collaboration), "Observation of a New Boson at a Mass of 125 GeV with the CMS Experiment at the LHC," *Phys. Lett. B* 716 (2012) 30.

[31] Tanabashi, Particle Physics, 94.

[32] Tanabashi, Particle Physics, 41.

[33] D. Gross, and F. Wilczek, *Phys. Rev. Lett.* 18 (1973): 1174; D. Politzer, *Phys. Rev. Lett.* 30 (1973): 1346; S. Weinberg, *Phys. Rev. Lett.* 31 (1973): 494; H. Fritzsch, M. Gell-Mann and H. Leutwyler, "Advantages of the Color Octet Gluon Picture," *Phys. Lett. B* 47 (1973): 365.

[34] M. Gell-Mann and Y. Ne'eman, *The Eightfold Way* (New York: Benjamin, 1964).

[35] Gell-Mann, Schematic Model.

[36] Ibid.

[37] YouTube, https://www.youtube.com/watch?v=V0kZLSai4tc .

[38] Bodek, Structure of the Nucleon.

[39] E.D. Bloom, *et al.*, "High Energy Inelastic *e-p* Scattering at 6° and 10°," *Phys. Rev. Lett.*, 20, 16 (1969): 930.

[40] E.E. Chambers, and R. Hofstadter, "The Structure of the Proton," HEPL-78, Stanford University, April 1956.

[41] J.D. Bjorken, "Asymptotic sum rules at infinite momentum," *Phys. Rev.* 179 (1969): 1547.

[42] C.G. Callan and D.J. Gross, "High-energy electroproduction and the constitution of the electric current," *Phys. Rev. Lett.* 22 (1969): 156.

[43] J.D. Bjorken and E.A. Paschos, "Inelastic Electron-Proton and γ-Proton Scattering and the Structure of the Nucleon," *Phys. Rev.* 185 (1969): 1975.

[44] R.P. Feynman, "Structure of the proton," *Science* 183 (1974): 601; R.P. Feynman, "Very high-energy collisions of hadrons," *Phys. Rev. Lett.* 23 (1969): 1415.

[45] D. Drell, *et al.*, "Theory of Deep-Inelastic Lepton-Nucleon Scattering and Lepton Pair Annihilation Processes II, Deep-Inelastic Electron Scattering," *Phys. Rev. D* 1 (1970): 1035.

[46] Sidney D Drell and Tung-Mow Yan, "Partons and their applications at high energies," *Annals of Physics* 66 (1971): 2.

[47] A. Bodek, *et al.*, "Experimental studies of the neutron and proton electromagnetic structure functions," *Phys Rev D* 20 (1979): 1471.

[48] J. Kuti and V.F. Weisskopf, "Inelastic Lepton-Nucleon Scattering and Lepton Pair Production in Relativistic Quark-Parton Model," *Phys. Rev. D* 4 3418 (1971): 3419.

[49] V. Tvaskis, *et al.*, "Proton and deuteron F_2 structure functions at low Q^2," *Phys. Rev. C*, 81, (2010): 305.

References

50 The mass of a proton is 8.88 times that of a muon. The mass of the proton is 1,836.15 electron masses, compared to 206.77 electron masses for the muon (P.J. Mohr, *et al.* "CODATA Recommended Values of Fundamental Physical Constants: 2014," arXiv:1507.07956v1 [physics.atom-ph]).

51 T. Ahmed *et al.* (H1 Collab.), "Observation of deep inelastic scattering at low x," *Phys. Lett. B* 299 (1993): 385.

52 I. Abt *et al.* (H1 Collab.), "Measurement of the proton structure function F_2 (x, Q^2) in the low-x region at HERA" *Nucl. Phys. B* 407 (1993): 515.

53 J. Feltesse, "HERA the new frontier," Presented at Conference: C91-08-05 (SLAC Summer Inst. 1991: 155-205), p.155-205, CEA-DAPNIA-SPP--92-01.

54 T. Ahmed *et al.* (H1 Collab.), "A measurement of the proton structure function $F_2(x, Q^2)$," *Nucl. Phys. B* 439 (1995): 47.

55 V.Z. Peterson, "Mesons produced in proton-proton collisions," University of California Radiation Laboratory, URCL-713 (1950).

56 Martin Spergel, Michael Lieber and S. Milford, "Pion production in high-energy cosmic-ray collisions," *Il Nuovo Cimento A* 42 (1966): 251.

57 R. Shyam, W. Cassing and U. Mosel, "Exclusive pion production in proton-nucleus collisions and the relativistic two nucleon dynamics," *Nucl. Phys. A* 586 (1995): 557.

58 D. B. Lichtenberg, "Pions in proton-proton collisions," *Phys. Rev.* 100 (1955): 303; C. Hanhart, J. Haidenbauer and J. Speth, "Pions in proton-proton collisions," *Acta Physica Polonica B* 29 (1998): 3047.

59 W. Huang and L.Yu "Serial Symmetrical Relocation Algorithm for Equal Sphere Packing Problem," arXiv:1202.4149v1 [cs.DM] (2012).

60 Tanabashi, Particle Physics, 98.

61 Tanabashi, Particle Physics, 99.

62 Tanabashi, Particle Physics, 108.

63 Tanabashi, Particle Physics, 50.

64 Tanabashi, Particle Physics, 58.

65 Tanabashi, Particle Physics, 64.

66 Tanabashi, Particle Physics, 36.

67 Ibid.

68 Ibid.

69 Tanabashi, Particle Physics,56.

70 Tanabashi, Particle Physics, 36.

71 A.D. Polyanin and A.I. Chernoutsan, *A concise handbook of mathematics, physics, and engineering sciences*, 430, (Boca Raton: Chapman & Hall/CRC, 2011).

72 Tanabashi, Particle Physics, 36.

73 Tanabashi, Particle Physics, 37.

74 Tanabashi, Particle Physics, 47.

75 Tanabashi, Particle Physics, 50.

76 Tanabashi, Particle Physics, 56.

77 W. Pauli, "Open letter to the group of radioactive people at the Gauverein meeting in Tübingen," Zürich, Dec. 4, 1930.

78 E. Fermi, "Tentativo di una teoria dei raggi beta" (Attempt at a theory of beta rays), *Il Nuovo Cimento* 9 (1934): 1.

79 J. Schwinger, "A Theory of the Fundamental Interactions," *Annals Phys.* 2 (1957): 407.

80 T.D. Lee, "Intermediate boson hypothesis of weak interactions," Proceedings, 10th International Conference on High-Energy Physics (ICHEP 60): Rochester, NY, USA, 567, 1960.

[81] G. Arnison, *et al.*, (UA1 Collaboration) "Experimental observation of isolated large transverse energy electrons with associated missing energy at s = 540 GeV," *Phys. Lett.* 122B (1983): 103.

[82] Tanabashi, Particle Physics, 33

[83] Tanabashi, Particle Physics, 34.

[84] Tanabashi, Particle Physics, 64.

[85] Peter W. Higgs, "Broken Symmetries and the Masses of Gauge Bosons," *Phys. Rev. Lett.* 15 (1964): 508.

[86] G. Aad, *et al.*, "Observation of a new particle in the search for the Standard Model Higgs boson with the ATLAS detector at the LHC," *Phys. Lett. B* 716 (2012): 1.

[87] G.S. Guralnik, *et al.*, "Global Conservation Laws and Massless Particles," *Phys. Rev. Lett.* 13 (1964): 585.

[88] F. Englert and R. Brout, "Broken Symmetry and the Mass of Gauge Vector Mesons," *Phys. Rev. Lett.* 13 (1964): 321.

[89] J. Bahcall and G. Shaviv, "Solar Models and Neutrino Fluxes," *Ap. J.* 153 (1968): 113.

[90] J. Bahcall, "An Introduction to Solar Neutrino Research," arXiv:hep-th/9711358v1 (1997)

[91] J. Bahcall, "Solar Neutrinos. I. Theoretical," *Phys. Rev. Lett.* 12 (1964): 300.

[92] B.T. Cleveland, *et al.*, "Measurement of the Solar Electron Neutrino Flux with the Homestake Chlorine Detector," *Ap. J.* 496 (1998): 505.

[93] J. Bahcall, "Solar neutrinos: an overview," *Phys. Rep.* 333-334 (2000): 47.

[94] R. Davis, "Report on Solar Neutrino Experiment," BNL-31974 (1982).

[95] Cleveland, Solar Electron Neutrino Flux.

[96] J. Kiko, (GALLEX Collaboration), "The GALLEX solar neutrino experiment at the Gran Sasso Underground Laboratory," *Astrophysics and Space Science* 228 (1995): 107.

[97] M. Altman, *et al.*, (GNO Collaboration), "Complete Results for Five Years of GNO Solar Neutrino Observation," *Phys. Lett. B* 616 (2005): 174.

[98] J. Bahcall, "Gallium solar neutrino experiments: Absorption cross sections, neutrino spectra, and predicted event results," *Phys. Rev. C* 56 6 (1997): 3391.

[99] J.N. Abdurashitov, *et al.*, (SAGE Collaboration), "Measurement of the Solar Neutrino Capture Rate with Gallium Metal," *Phys. Rev. C* 60 5 (1999): 055801.

[100] B. Pontecorvo, "Mesonium and antimesonium," *J. Exptl. Theoret. Phys.* 33, (1957): 549.

[101] B. Pontecorvo, "Neutrino experiments and the question of leptonic-charge conservation," . *Exptl. Theoret. Phys.* 26 (1968): 984.

[102] Y. Suzuki, "The Super-Kamiokande Experiment," *Eur. Phys. J. C* 79 (2019): 298.

[103] Y. Fukuda, *et al.*, (Super-Kamiokande Collaboration), "Evidence for Oscillation of Atmospheric Neutrinos," *Phys. Rev. Lett.* 81 (1998): 1562.

[104] A. Bellerive, *et al.*, (SNO Collaboration), "The Sudbury Neutrino Observatory," *Nucl. Phys. B* 908 (2016): 30.

[105] Q.R. Ahmad, *et al.*, (SNO Collaboration), "Direct Evidence for Neutrino Flavor Transformation from Neutral-Current Interactions in the Sudbury Neutrino Observatory," *Phys. Rev. Lett.* 89 (2002): 011301.

[106] Bahcall, Solar neutrinos.

[107] *Astronomy & Geophysics*, Volume 45, Issue 4, August 2004, Pages 4.21–4.25.

[108] K. Lande, "The Life of Raymond Davis, Jr. and the Beginning of Neutrino Astronomy," *Annu. Rev. Nucl. Part. Sci.* 59 (2009): 33.

[109] R. Davis and J.C. Evans Jr., "Report on the Brookhaven Solar Neutrino Experiment," BNL-21837 (1976).

Index

Index

N

Index

all the same, 114
charge, 4
electron same as muon?, 113
existence verification, 4
flavors
    electron (ν_e), 4
    muon (ν_μ), 4
    tau (ν_τ), 4
fundamental particle, 118
hypothesized, 75
interaction with matter, 128
mass, 4, 86
    dilation, 114
    electron equal to muon, 113
    from defect equation, 112, 113
    negative, 112, 113
one for all charged leptons, 116
only one type, 115
oscillations, 88, 90, 93, 114
pair production, 121, 125, 128
produced in Sun, 88
solar neutrino problem, 88
universal neutrino, 116
veiled charge, 126
neutrons
    as unitary triplets, 10
    decay, 75
        internal electron-positron
            annihilation, 79
        Standard Model, 79
            diagram, 79
        W boson emission, 79
        W^- emission, 76
            diagram, 76
    discovery, 1
    properties, 5
    structure
        pions, 28
        quarks
            not components, 29
non-integral charges, 10

P

parallel axis theorem, 54
particle jets, 83
partons
    hypothesized, 11

theory, 11, 15
Pauli, Wolfgang, 75
periodic table, 1, 2
pions
    as decay products, 32
    charged (π^+, π^-), 3
        decay, 34, 39, 61, 78, 109
            electron-positron annihilation,
                34, 61
            muon neutrino emission
                Standard Model, 35
            muon neutrino pair
                production, 35, 62
            Standard Model, 78
                diagram, 79
            W boson emission, 78
        mass, 34, 39, 109
            beta vs. free electrons, 110
        model advantages, 62
        possible model, 36
        structure, 34, 36, 39
            central cluster of betas, 61
            diagram, 62
            electron-positron, 34, 111
            electron-positron shell, 62
            valence electron, 34
            valence positron, 34
    decay, 6, 7
        electron-positron annihilation, 47
        neutrino pair production, 47
        π^- (Standard Model), 33
    discovery, 2
    neutral (π^0)
        decay, 34, 63, 111
            electron-positron annihilation,
                34
            into electron-positron pair, 63
            into electron-positron pairs,
                110
            into photons, 110
        mass, 34, 63, 109, 111
            beta vs. free electrons, 110
        possible model, 36
        structure, 34, 36, 39
            central cluster of betas, 63
            diagram, 63
            electron-positron, 34, 111
    structural unit, 39
        baryons, 33, 85

Q

Index

ABOUT THE AUTHOR

William Stubbs is a retired engineer who independently researches a variety of subjects including nuclear and particle physics. He earned a degree in Nuclear Engineering from the University of Tennessee and worked for both private and public engineering organizations during his 30-year career. His former employers include General Electric, Westinghouse Electric, the Tennessee Valley Authority, and the U.S. Department of Energy.

William has published several articles on physics and nuclear science since retiring in 2005, and self-published five other books: *Nuclear Alternative: Redesigning Our Model of the Structure of Matter* in 2008; *Gravity* in 2012; *Proton Structure* in 2015; *The Physics of Bodies in Real and Imaginary Spaces* in 2016; and *The Nucleus of Atoms: One Interpretation* in 2018.

In addition to his research, William enjoys listening to, performing, and composing music; watching classis sci-fi movies, and spending time with family and friends. He lives in Port St. Lucie, Florida. Comments and correspondence may be sent to ift22c@bellsouth.net.

www.ingramcontent.com/pod-product-compliance
Lightning Source LLC
Chambersburg PA
CBHW030938240526
45463CB00015B/387